The Patrick Moore Practical A

More information about this series at http://www.springer.com/series/3192

Astronomy for Older Eyes

A Guide for Aging Backyard Astronomers

James L. Chen

Graphics by Adam Chen

James L. Chen
The Shenandoah Astronomical Society
Gore, VA, USA

Adam Chen
Baltimore, MD, USA

ISSN 1431-9756 ISSN 2197-6562 (electronic)
The Patrick Moore Practical Astronomy Series
ISBN 978-3-319-52412-2 ISBN 978-3-319-52413-9 (eBook)
DOI 10.1007/978-3-319-52413-9

Library of Congress Control Number: 2017931679

Cover Illustration: Helix Nebulae taken by NASA's Spitzer Space Telescope

Printed on acid-free paper

This Springer imprint is published by Springer Nature
The registered company is Springer International Publishing AG
The registered company address is: Gewerbestrasse 11, 6330 Cham, Switzerland

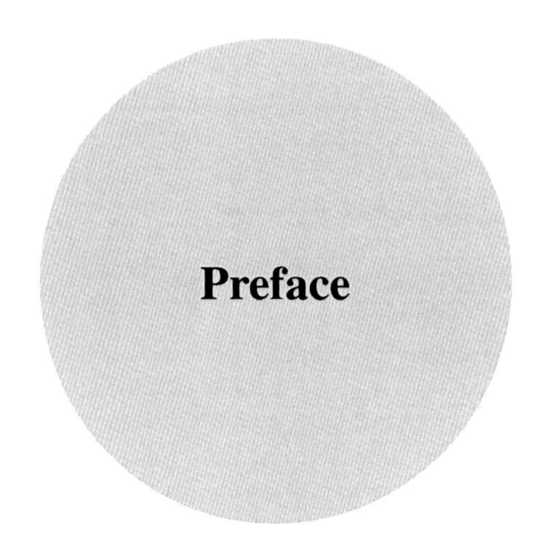

We are stardust,
We are golden,
We are billion year old carbon,
And we got to get ourselves back to the garden.

—"Woodstock" by Joni Mitchell

I'm getting old.

I don't know how it happened.

At one time, I was young, physically fit, mentally active, and building a career in engineering. Then I got married. Had kids. My wife and I raised two fine boys who have become two fine men. I retired early at age 52, having worked 32 solid and productive years for the federal government. My post-retirement life has found me working as an engineering consultant, a swimming coach, and guitar teacher, and now as an author.

My parents gave me my first telescope for Christmas when I was 11 years old. Many telescopes have passed through my hands, as backyard astronomy has been a passion for over 50+ years.

In my 55th year, the first signs of my growing old began to reveal themselves. Reading books became arduous as my eyes would grow weary. My ophthalmologist told me I needed bifocals. Strike one.

My physician told me I had high blood pressure and cholesterol. I had to start taking pills every day. My doctor told me I had to lose weight. No more fried chicken, apple pie a la mode, french fries, potato chips, cheese nachos, and other greasy gut-bomb foods that are so comforting. Strike two.

Strike three came on a crisp October evening. Unknown to me, my physical strength was slowly waning. The sky was crystal clear. There was a faint crescent Moon. A perfect night for my big 11" Schmidt-Cassegrain, weighing in at a hefty 92 lb including tripod.

I traded for this behemoth while suffering an extreme case of aperture fever. A 3.5" Questar for a Celestron 11 GPS, a straight up deal. And I have never regretted that deal. As beautifully made as the Questar is, it can never show the deep sky delights that 11 inches of aperture can provide.

This big instrument is the most fun telescope I have even owned, with its full GoTo capability and ample aperture. But it weighs 92 pounds!

In my younger days, moving this behemoth from the living room onto the deck of my house was easily accomplished. On this night, the 11 GPS seemed to have gained weight. It felt especially heavy. I maneuvered the telescope through the door onto the deck with a degree of difficulty that I hadn't experienced previously. As I struggled to set the telescope down into its usual position on the deck, one of the legs of the tripod tangled with a leg of the picnic table. I stumbled. I fumbled. I dropped the telescope onto the deck! I stood speechless and in shock over my telescope in horror. My wife came out to see what was wrong, having heard the loud crash of the scope hitting the deck from the den. I was still frozen, standing over my beloved scope, unable to comprehend what had just happened.

Fortunately, the only casualty of the incident was a cracked aluminum corrector plate cell on the telescope. The Schmidt corrector and the secondary mirror were intact, and in fact the secondary was still in alignment. Lucky me. A quick call to Celestron the next day and $100 later, I was able to obtain a new corrector plate cell and repair the telescope. Whew! (Figs. 1, 2, and 3).

Fig. 1 The author's Celestron 11 GPS

Fig. 2 The cracked Schmidt corrector plate cell

Fig. 3 Another angle of the cracked Schmidt corrector plate cell

This accident taught me a lesson. I had to start making adjustments if I wanted to continue to pursue astronomy as a lifelong passion.

My eyes aren't what they used to be. I wasn't as strong as I used to be. My sense of balance is starting to be questionable. Cold nights seemed colder. Warm nights seemed warmer and sweatier. Mosquitoes seemed to have found me tastier. And the simple act of bending over to peer through the eyepiece was becoming problematic due to a somewhat bulbous belly that had started to form about my midriff.

Two years ago, I complained to my ophthalmologist that I thought I needed new glasses. He asked how did I know. I told him when I played golf (another passion of mine!), I couldn't see the flight of my golf ball until it splashed into the water hazard! Haha! Unfortunately, his examination revealed that at age 62, I had the beginnings of cataracts. Strike four! (wait a minute, that's not right....)

How am I going to survive this aging process and still pursue my passion for astronomy? Thus, the genesis that led to writing this book.

Another theme contained in this book is the need for older people to socialize, and astronomy clubs and star parties provide socialization opportunities.

My involvement with my astronomy club, the Shenandoah Astronomical Society, increased my interaction with the public. One of the main goals of my club is public outreach, especially reaching out to introduce astronomy to young people. When did young people become so young? When did I get so .... uh..... distinguished?!? Many of my friends in the club are like me, retired and still in love with astronomy. Nothing gives us a greater thrill than having a young girl or boy peer through our telescopes and exclaim "WOW!"

It reminds me of one extraordinary night, when I was in my thirties and my older son was a mere four years old. I was out on my back deck of our house, with my trusty Celestron C-5 telescope set up and viewing the Great Orion Nebula.

My first born opened up the sliding glass door and asked "What are you doing, Daddy?"

"I'm looking through my telescope at the Great Orion Nebula," I replied.

"Can I see?" he asked.

Thus began a wondrous evening with my young son. Propping him up, he got his first views of M42, Mizar and Alcor in Ursa Major, and the giant planet Jupiter through my telescope. The viewing session ended as a slow moving cloud front creeped in and eventually obscured the night sky.

But my son's enthusiasm and curiosity about astronomy was forever born that night, as he and I brought in my telescope equipment and spent the rest of the evening looking at the pictures of galaxies in Timothy Ferris' *Galaxies* and back issues of *Astronomy* and *Sky and Telescope* magazines.

My son has grown up to be a fine man. What happened to his interest in astronomy? Look at the covers of all of my books. Adam Chen has been the graphics designer for every book of mine. He has worked as a support contractor for NASA. He has helped me on all my books by providing drawings, illustrations, graphics, and cover layouts, and most importantly inspiration.

Another favorite memory of mine, involving my younger son. When Alex was four years old (there seems to be a pattern here!), I took him on an across-the-country journey to attend the Riverside Telescope Makers Convention, known as RTMC, near Big Bear Lake in the San Bernardino Mountains of California. We flew into Ontario Airport and I drove us to Big Bear Lake in a rental car. The first night at RTMC, my son and I were treated to a view of the Whale Galaxy (NGC 4631) and the Hockey Stick Galaxy (NGC 4656) through a magnificent 24-in.

Dobsonian telescope set up near the center of the telescope field. We talked with many fellow RTMC attendees that night, and saw many other celestial wonders through other telescopes. My son was absolutely thrilled as the first night drew to a close on this astronomy adventure.

The next morning, as my son and I marched up the hill to attend the vendor sales and swap meet that RTMC schedules on Saturday morning, a very unusual weather event took place. On Memorial Day weekend, the unofficial beginning of the summertime season, snowflakes fell. It was snowing at RTMC! My son didn't care, this was an adventure and he was his Dad! After RTMC, I drove my son down to Anaheim and took him for a day at Disneyland before we returned to the airport to fly home.

My younger son has grown up to be a fine young man also. Through his years growing up, he and I attended many more star parties together. He went to the Ohio State University and earned a bachelor's degree in aeronautical engineering and is now completing his master's thesis. He fell in love with my 4-in. brass Renaissance telescope, so on his sixteenth birthday, I gave him the telescope as a gift. Alex Chen uses "Brassy" to this day.

If you read my biography, you will notice that I have spent parts of three decades working part time at two different telescope stores in the Washington, D.C. area. There are many people who I have met and have purchased equipment from me at the North East Astronomy Forum over the past 15 years. Many of my insights and recommendations on telescope equipment are drawn from my experience working with both experienced amateur astronomers and people new to the astronomy hobby. Two of my previous books, *The Vixen Star Book User Guide* and *The NexStar Evolution and SkyPortal User Guide* are the result of my expertise gained from selling and teaching the use of this equipment to customers.

I expect some of my equipment recommendations in this book will cause some controversy and start discussions amongst my peers. That's a good thing. The one thing I have discovered over the years is that there is no singular answer that applies to everybody.

This is the story of how I came to write this book. Aging is inevitable. But with the right lifestyle adjustments, this wonderful activity of backyard astronomy is survivable. I will continue to make memories with my sons and my friends and to show the next generation the wonders of the universe from a backyard telescope.

Gore, VA
August, 2016

James L. Chen

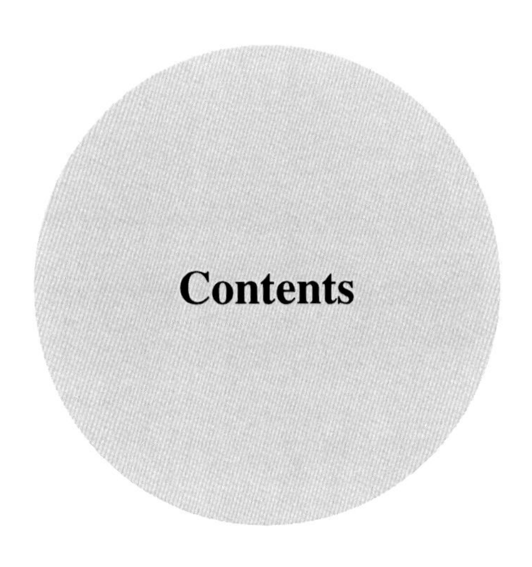

Contents

About the Authors

James L. Chen is retired from the Department of the Navy and Federal Aviation Administration where he worked as a Radar and Surveillance Systems Engineer. A guest lecturer at local Washington, DC/Northern Virginia/Maryland astronomy clubs on amateur astronomy topics of eyepiece design, optical filters, urban and suburban astronomy, and lunar observing, he wrote an *Astronomy Magazine* article on Dobsonian telescope design in the November 1989 issue. His first book was published in June 2014 by Springer, entitled *How to Find the Apollo Landing Sites*. His second book entitled *A Guide to the Hubble Space Telescope Objects* is also available from Springer. His third and fourth books are user guides for GoTo telescopes and mounts: *The Vixen Star Book User Guide* and *The NexStar Evolution and SkyPortal User Guide.* He served as a part-time technical and sales consultant to two Washington DC area telescope stores for over 30 years.

Adam Chen is former Program Manager of media support for NASA Headquarters in Washington DC. He is creator and executive producer of major NASA publications, including the book and web-book application documenting the history of the Space Shuttle Program "Celebrating 30 Years of the Space Shuttle Program." Currently he works in marketing for Brown Advisory, an investment firm in Baltimore, MD. Adam Chen has been the graphic designer for all of James L. Chen's books.

Chapter 1

Amateur Astronomy and Its Aging Practitioners

I've loved the stars too fondly
To be fearful of the night.

—*Galileo*

There are two types of people in the world. Young people and people who used to be young.

There are two types of people who are interested in astronomy. People who have been observing through telescopes for years and those who always wanted to look through a telescope at the night sky.

Every year, countless astronomy clubs all over the world hold public outreach events and star parties, inviting both young people and older people to have their first look through a telescope to view the wonders of the night sky. How lucky these people are, to behold the awesome treasures normally hidden by light pollution and by the object's faintness and distance, treasures that can be only to be revealed with a telescope.

Many of today's amateur astronomers entered the hobby in their teenage years during the late 1950s or 1960s. This was the heyday of the Space Race, where the world's two great superpowers engaged in scientific and engineering fisticuffs in an effort to show the world whose technology and politics was superior. This was the Golden Age of Amateur Astronomy. The magazine *Sky and Telescope* became the monthly bible, where readers found out the latest scientific discoveries, the monthly celestial phenomena worth observing, and the plans of a telescope maker's latest creation. Amateurs were grinding their own mirrors and building telescope mounts

J.L. Chen, *Astronomy for Older Eyes*, The Patrick Moore Practical Astronomy Series, DOI 10.1007/978-3-319-52413-9_1

Fig. 1.1 The author's SCT and refractors (James Chen)

out of plumbing supplies. Amateurs would send snail mail orders to obtain eyepieces, optical rouge, and focusers.

The 1970s ushered in the rise in availability of commercial telescopes and their accessories, led by the introduction of mass-produced Schmidt–Cassegrain telescopes. *Astronomy* magazine entered the publishing scene as an alternative to *Sky and Telescope.* The availability of quality telescopes at affordable prices expanded the amateur astronomy ranks with a new population of sky watchers. The popularity of homemade telescopes began to wane (Fig. 1.1).

The early 1980s ushered in a new generation of amateur telescope makers with the teachings of a former monk named John Dobson. Newtonian telescopes returned to the telescope scene, with simple wooden mounted Dobsonian telescopes proliferating star parties and backyards. Telescope makers were drawn to the simple engineering and large apertures that the design offered. Homemade telescopes, using Dobson concepts, became fashionable again (Fig. 1.2).

The late 1980s and 1990s completed the telescope picture with the introduction of high-technology computer-driven telescopes and the reintroduction of refractors to amateur astronomy. Those new to the hobby entered a confusing world of alternative choices in the selection of their telescope optics and mounts.

In the ultimate mating of two hobbies, computers and astronomy, computer-controlled telescopes captured the backyard astronomer's imagination and pocketbook. Known collectively as GoTo telescopes, this advanced technology is fascinating to watch as the mount proceeds to point the telescope from object to object with precision, accompanied with the sounds of motors whirring and gears

Fig. 1.2 The Dobsonian telescope (Hands-on-Optics archive)

meshing. The era of computerized GoTo telescopes began in 1984. Computer-controlled telescopes took form during the same period as the development of personal computers. During the 1980s, the US telescope company Celestron formed a business relationship with Vixen Company, Ltd. of Japan. The American company featured its home-grown Schmidt–Cassegrain telescope mounted on a computer-controlled equatorial mount from Vixen, and marketed them under the Celestron brand. Vixen of Japan developed the Sky Sensor, an economical system consisting of a Go To computer control system with motors designed to attach onto their portable German equatorial mount known as the Super Polaris.

These GoTo telescope mounts are wonderful pieces of technology. The GoTo technology allows for more efficient use of observing time by quickly finding objects in the night sky. Built into the hand controller is a microprocessor, firmware, and built-in memory catalog of the positions of thousands of stars, galaxies, nebulae, open star clusters, globular clusters, planetary nebulae, our solar system planets, and the Moon. Complex algorithms developed and refined over years with improvements in encoders and motor technology have made the GoTo telescope an

Fig. 1.3 Celestron NexStar Evolution 6 GoTo Telescope (James and Adam Chen/Celestron composite photo)

accepted and desirable telescope feature. Computer-controlled telescopes can help its owner to overcome the fear of looking ridiculous while others watch; no longer will the telescope owner appear incompetent as he tries to find celestial wonders—now he only looks ridiculous as he tries to remember how to set up his telescope! (Fig. 1.3).

Concurrently, a revolution in the late 1980s and 1990s in refractor design resulted in the availability of high-priced, high-quality achromatic and apochromatic refractors. Using newly available exotic low dispersion glasses and innovative two lens, three lens, and sometimes four lens designs, another new group of telescope users were welcomed into the astronomy hobby. These new refractor designs offered high-contrast and high-quality stellar images, albeit at high cost. The demand for these Ferrari's and Lamborghini's of the telescope world often resulted in consumers spending time on wait lists as long as 2 or 3 years. Those looking for a profit sometimes took delivery of their long-awaited apochromatic refractors and immediately turned around and sold them for twice the original cost to those who didn't want to wait (Fig. 1.4).

Fig. 1.4 The author's apochromatic refractors (James Chen)

As amateur astronomy entered the new millennium, a downward trend became evident. The hobby was becoming progressively older. Those in the hobby tended to stay in the hobby. But younger people were not attracted to sitting in the night air peering through a telescope. The fascination of the Internet, computer gaming, and social media was far more powerful to the young than any telescope.

This downward trend became common among many traditional pastimes. Stamp collecting, coin collecting, model building, model railroading, and ham radio are all suffering from a lack of interest from young people. And with the proliferation of science fiction and fantasy movies and computer gaming, many young people find the quiet serenity of sitting at a telescope's eyepiece as too tame and lacking of action.

To add to the youth disinterest, sci-fi movies and computer gaming also have created a false sense that space has already been conquered, and this is far from the truth. There is no warp drive, teleportation, or subspace communication. Mankind is still in its early stages of space exploration.

Fortunately, there has been an upsurge in interest in backyard astronomy from an unexpected age-group. Many retirees and soon-to-be-retirees are taking up astronomy as an activity. Many are satisfying a lifelong curiosity that began during the early days of the Space Age. Some have reached a financial security level where they can afford a luxury item such as a telescope. Some are nature lovers wanting to explore more of what the universe has to offer.

The low numbers of young people entering the hobby and the influx of retirees seeking new activities has resulted in amateur astronomy skewing towards the older age groups. Amateur astronomy has evolved from the Do-It-Yourself 30-somethings, to the affluent 40-or-50-somethings, to now the well-heeled 60+ aged pensioner seeking something to do.

Entering the world of amateur astronomy can be a daunting task. Many older people are uncomfortable with the techno-jargon of the hobby. The technology and cost of telescopes and telescope accessories can be off-putting. If one has no background in the hard sciences such as physics, mathematics, and chemistry, there is a discomfort with astronomy to the degree of avoiding the subject altogether.

Fortunately, the practitioners of amateur astronomy are a patient and welcoming lot. Astronomy clubs everywhere have open meetings and outreach programs to bring in newcomers. The education of the masses is a principle goal of both astronomy clubs and individual amateurs. There is no competition. Snobbery is kept to a minimum. Every amateur delights in showing people the views through their treasured telescopes. The greatest compliment an amateur astronomer hears when showing the Moon, planet, or nebula to a newcomer is to hear "Wow!," "That's incredible!," or "It's beautiful!"

Chapter 2

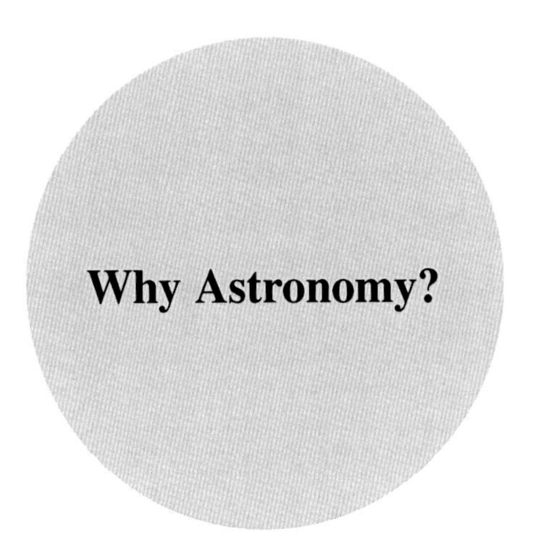

When I look up at the night sky,
and I know yes,
we are part of this universe,
we are in this universe,
but perhaps more important than both of these facts,
is that the universe is in us.
When I reflect on that fact,
I feel big.

—Neil deGrasse Tyson

Astronomy is a science that has fascinated and captivated mankind since the beginning of man's history. Mankind has looked to the heavens and pondered his place in the universe. Astronomy has impacted mankind's art, literature, and beliefs throughout the centuries.

For those who live in a city or brightly lit suburb, the night sky is often hidden by the light pollution caused by streetlights, parking lot lights, neon-lit business signs, and even front porch lights. But a short car ride to the country or to a dimly lit suburb, and the night sky reveals its wonders.

Astronomy is a natural science which studies stars, galaxies, planets, moons, asteroids, comets, and nebulae, and the evolution of celestial objects. Astronomy also includes observing and studying phenomena that originate outside the atmosphere of Earth, including supernovae explosions, gamma ray bursts, and cosmic microwave background radiation. A subset of astronomy is cosmology, which is concerned with studying the universe and its beginnings and evolution. Planetary

J.L. Chen, *Astronomy for Older Eyes*, The Patrick Moore Practical Astronomy Series, DOI 10.1007/978-3-319-52413-9_2

science studies the planets, moons, dwarf planets, comets, asteroids, other bodies orbiting the Sun, and exoplanets orbiting distant stars.

People of all ages can participate in amateur astronomy, and gain an understanding and well-being of the universe. This is particularly true for those recently retired individuals seeking a new activity or challenge in their life. Backyard amateur astronomy enables an individual to gain an understanding of objects and phenomena outside of Earth's atmosphere. Astronomy has always been and continues to be that rare activity where an amateur can share contributions with professionals in advancing the scientific knowledge of astronomy.

Many people envision astronomers as nerdy geeks (Okay, some of us are nerdy!) who stare through long white telescopes on isolated hills, but the truth is that there are thousands of amateur astronomers in the world who enjoy the hobby from their backyards. These backyard astronomers come from all walks of life, with both blue-collar workers and white-collar professionals participating. Amateur astronomy welcomes people from all levels of education, from high school level to post-graduate degrees (Fig. 2.1).

Astronomy involves observing and patience. In this often hectic and competitive world, astronomy offers a calmness where the serenity of the night sky can bring a peace of mind. A Zen-like oneness with the universe can permeate one's soul, as an observer can often find themselves in a meditative state under a clear night sky. Astronomy is a most fulfilling and rewarding hobby to pursue. Astronomy as a hobby is something that can be shared with friends, or can be enjoyed solo.

Fig. 2.1 Historic photo of Lord Ross's The Leviathan of Parsonstown (William Parsons)

There is a misconception that astronomy requires sophisticated, expensive equipment. However, many begin in the hobby just performing naked-eye observations, or using a pair of binoculars. Observing meteor showers, identifying constellations, and seeing man-made objects, like the International Space Station as it passes overhead, are activities that require no special equipment for beginners. Many find great satisfaction scanning the night sky with binoculars, finding a treasure trove of star clusters, a few galaxies, planets, and details of the Moon visible through binoculars. As one progresses in the hobby, a telescope becomes a tool for conducting further detailed observations and study, and depending on economics, there are telescopes available for all budgets. Even the simplest, most affordable telescope will reveal celestial wonders to the observer's eye. Of course, for the well-heeled, the sky's the limit (pardon the pun!). For those less fortunate, public observatories and public outreach programs abound in the United States, many supported by astronomy clubs peopled with enthusiastic amateurs willing to show people the beauty and wonders of the night sky. In fact, many astronomy clubs and some public libraries offer loaner telescopes to the public that can be checked, used, and returned like a library book!

Choosing What Kind of Backyard Astronomer?

Being an amateur backyard astronomer is not just one single pursuit. There are many subcategories associated to being an amateur astronomer (Fig. 2.2).

Fig. 2.2 The author's early photo of the Moon (James Chen)

Fig. 2.3 William Parsons, The Third Earl of Rosse drawing of M51 (William Parsons)

Self-Discovery/Historical Astronomy—For most people, the first stage of becoming an amateur astronomer is part self-discovery and part appreciation of the celestial beauty. Often, when a new galaxy, nebula, or star cluster is viewed for the first time, all astronomers share the same emotions of discovery and awe that the pioneering astronomers felt. When searching and finding a deep sky object, the observer is following the historical footsteps and observations of Galileo, Sir John Herschel, Charles Messier, Johann Elert Bode, Pierre Mechain, Admiral William Henry Smyth, and numerous other historical astronomy figures. When an object is seen for the first time by an amateur, it is their first time to view it, just as it was for the historical astronomers. It is a thrill that is both personal, yet shared with many astronomers (Fig. 2.3).

Scientific Astronomy—There are amateur astronomers that participate in actual scientific studies and contribute to the science. There are devout amateurs who perform occultation timings, where an asteroid passes in front of a star and blocks the star's light. Numerous occultation timings are performed across the country, with the timing data of the occultation being sent to the occultation organization for processing, resulting in a determination of the asteroid's shape and size. There are amateurs that regularly make variable star observations, providing data on variations of brightness of variable stars. This data is then passed on to observatories or universities to compile and analyze. In the past, comet and asteroid hunting was left up to the amateur astronomy community, although that has now being taken over by professional astronomers seeking to avoid worldwide destruction from a comet or asteroid striking the Earth. Initial supernova observations are often reported by amateur astronomers and reported to observatories. Astronomy remains one of the few sciences where an amateur can make a contribution.

Fig. 2.4 An amateur astrophoto of M45 The Pleiades (John Livermore)

There are programs that exist where nonscientific members of the public contribute to data reduction or data analysis to aid observatories. For example, high school students in Northern Virginia and West Virginia regularly analyze radio telescope data from the National Radio Astronomy Observatory at Green Bank, West Virginia, to discover new pulsars. A number of new discoveries have been identified from this program, contributing to the science of astronomy and adding to the higher education of scientifically minded youth (Fig. 2.4).

Astrophotography—Countless number of people involved with astronomy want to take pictures of their observations. There is a technological desire of those first entering the hobby to share their observations in the form of images of the stars, galaxies, planets, etc. In this day of point-and-shoot cameras and cellphone cameras, the assumption is that astrophotography is easy. Nothing can be further from the truth. Any previous experience in photography is minimized when attempting to image the celestial sky. Whether using a dedicated CCD astro-camera, a modified DSLR, or doing it the old-fashioned way using a film camera, there is a steep learning curve that must be climbed by the novice backyard astronomer. This activity is not for the faint-of-heart. The backrooms of many telescope stores are filled with telescopes, equatorial mounts, and camera adapters from failed attempts at this part of the astronomy hobby. Be forewarned. Please spend 2 or 3 years in naked-eye observing before entering the minefield known as astrophotography. That way, if astrophotography isn't appealing, the hobby of astronomy will still be there (Fig. 2.5).

Public Outreach and Space Education—Amateur astronomers and their astronomy clubs are dedicated to educating the public on the subjects of astronomy, space

Fig. 2.5 An amateur astrophoto of Jupiter (Jon Talbot)

sciences, and space exploration. Star parties are held at locations all over the United States and the world to serve as a common ground for both established observers and those new to astronomy to gather. Astronomy clubs all over the country hold public outreach sessions at local, state, and national parks inviting the public to see the night sky through the club members' telescopes. Educational talks about astronomy are given at public and private schools and libraries. Public education about astronomy is a very rewarding activity for many in the amateur astronomy community.

For this author, public outreach resulted in a teenage high school girl discovering the stars and eventually going to college to study astronomy. She is now working on her doctorate degree in astrophysics. All this because of three weekend nights when she and her parents came to the author's home to observe a comet through the author's collection of telescopes. She was so excited and enthusiastic about the comet and astronomy that she wanted a telescope to explore the night herself. Within weeks, her father and the author built a six-inch Dobsonian telescope in her father's garage. By her senior year in high school, she had become involved in occultation timings. At high school graduation, she was accepted to a prestigious California university to study astronomy. All this began because of three Saturday nights looking through a telescope. The positive impact on a young person's life can be significant and rewarding.

Gear Head—There are those in the hobby that are rightly categorized as Gear Heads. Telescopes, eyepieces, and accessories can be found in every room of their house.

As seen from the photos, this author is a part of the lunatic fringe known as gear heads. The author's excuse to his wife is as follows: "I don't drink. I don't smoke. I don't fool around. And you know exactly where I am at midnight, in the backyard with my telescope." (Figs. 2.6, 2.7 and 2.8).

Fig. 2.6 Part of the author's collection of telescopes (James Chen)

Fig. 2.7 More of the author's collection of telescopes (James Chen)

Fig. 2.8 The author's collection of eyepieces (James Chen)

As one gets more involved in astronomy, one discovers that one telescope doesn't do all jobs. Refractors are desirable for observing the Moon and planets. Newtonians (which includes the Dobsonian telescopes) are the best bang-for-the-buck big aperture telescopes. Catadioptrics, including Schmidt–Cassegrain and Maksutov designs, are highly portable. Eyepiece designs offer wide field, high contrast, or high eye relief for those who wear glasses. There are adapters for telescope mounts, and there are electronic drives, computerized mounts, camera adapters, and dew zappers. The list goes on and on. To some extent, every amateur becomes a gear head.

Gear Acquisition Syndrome, or GAS, is a common affliction among amateur astronomers. There are amateurs who collect telescopes with low serial numbers in an attempt to secure serial number 0001, or close to it. Some try to collect every eyepiece of a particular manufactured series or design. Often, an amateur will have a "big eye" telescope for deep sky observing and a "grab-and-go" telescope for situations as the name implies. Many planetary enthusiasts will insist on owning more than one high-resolution and high contrast refractor. In the era of GoTo telescopes, many amateurs are upgrading to computerized telescopes. The telescope manufacturers and telescope store owners are happy and more than willing to fulfill the GAS needs of the amateur astronomy populace.

Antique Restoration—There are a great number of aging and historic telescopes in the United States and the world awaiting restoration, and there is a small group of amateur astronomers who are devoting their time and energy in restoring these treasured scientific instruments to proper working order. Old telescopes exist in major observatories. Old telescopes exist in dark corners of garages or basements. There are dedicated and skilled amateurs who are willing to spend time and money

Fig. 2.9 A restored Alvan Clark 5-inch refractor at the U.S. Naval Observatory (USNO photo)

into cleaning and aligning the optics, rebuilding and lubricating mechanical parts, repainting and polishing telescope tubes and focusers, and making operational old telescope mounts. The smaller telescopes, such as an old Mogey 3-in. or Brashear 4-in. refractor, can be restored by a single individual in his workshop. Larger observatory based instruments are returned to operation with the efforts of a team of skilled technicians. The thrill for these people is the same for those who restore old automobiles or airplanes. The author has restored a Unitron/Polarex mechanical clock driven equatorial mount (Figs. 2.9 and 2.10).

Amateur Telescope Making—Many of today's amateur astronomers entered the hobby as amateur telescope makers during the late 1950s or 1960s. This was the Golden Age of Amateur Astronomy and the Golden Age of amateur telescope making. Each month *Sky and Telescope* magazine would feature plans for building a particular telescope. Amateurs were grinding their own mirrors and building telescope mounts out of plumbing supplies. These were the early days of astronomy mail order businesses, where eyepieces, optical rouge, mirror blanks, and focusers could be ordered. A second wave of amateur telescope making occurred during the 1980s spurred by the telescope making concepts of John Dobson (Fig. 2.11).

Even with the availability of excellent commercial telescopes and eyepieces, there is a contingent of telescope makers who love working with their hands to produce their own unique telescope. They take great pride and delight in observing the night skies with an optical instrument of their own design with optics ground and tested by themselves. With their own hands, they can build their own mounting

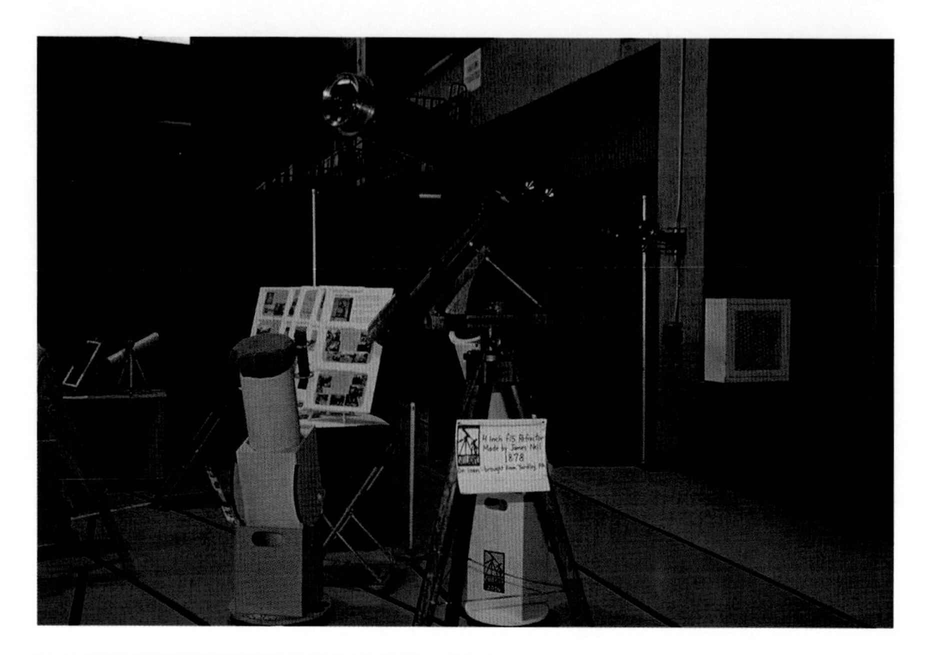

Fig. 2.10 A restored James Neil 4-inch f/15 refractor on display at NEAF 2016 (James Chen)

Fig. 2.11 A homemade DIY telescope (Hands-on-Optics photo archive)

Fig. 2.12 Vincent Van Gogh's The Starry Night (Van Gogh)

systems, and align their own optics. There are true craftsmen in the world of amateur astronomy.

Spiritual and Aesthetic Astronomy—There is an astronomy subculture who spend their nights gazing through their telescope, pondering their place in the universe, finding solace from the rapid pace of the world, and contemplating the greater glory of the world and heavens that surround them. To these people, backyard astronomy represents an opportunity to meditate and become one with the universe.

Some sky watchers are attracted to astronomy to observe Nature's beauty in the sky with its wide diversity. They appreciate the starkness of the Moon's craters, the ever changing clouds of Jupiter, the wisps of nebulosity of M42, the Great Orion Nebula, and the pinpoint jewels of light in M13, the Great Hercules Cluster.

Here is where the artists reside, peering into the night sky and drawing, painting, or writing of the wonders, the majesty, and the beauty of the universe. The world and the universe as seen through the artist's eye is much different than through an scientific eye. There is plenty of room in life for both viewpoints (Fig. 2.12).

Online/Remote Astronomy—Within the past few years, a number of Internet sites have been created to enable astronomers to use remote telescopes to observe. Each telescope is equipped with an imaging camera, and software applications allow the user to control the telescope from the comforts of home using their desktop/laptop, or tablet to observe. Although this is no substitute for actual eyeball-at-the-eyepiece observing, this is an interesting alternative to lugging telescope equipment outdoors and getting either frostbite or mosquito bites. This is a great opportunity for the handicapped astronomer to participate in the hobby. Examples of such sites are iTelescope.net, telescope.org, and skycenter.arizona.edu/programs/remote. Some computer savvy is required.

Bookshelf Astronomy—Pity the poor bookshelf astronomer, reading every science book at the library and from bookstores, yet never actually looking through a telescope. Like the related cousin, the bookshelf traveler who never travels, the bookshelf astronomer never experiences the joy, awe, and wonderment that peering through an eyepiece of a telescope can bring.

However, the bookshelf astronomer will gain knowledge of the historical past and the current state of the science that many backyard astronomers would envy. No matter which type of astronomy path that an individual chooses, there should be a little bookshelf astronomer mixed in.

Meteorite Hunting—An interesting offshoot of astronomy is that of a meteorite hunter. Part-astronomer, part-geologist, and a good helping of adventurer, this subset of amateur astronomy has a lot of appeal to the outdoorsmen. With the news of a fresh meteor fall or historical data about an old fall, the intrepid hunter will travel to often times remote locales, dig through mud, chop through high grass or trees to hunt down these rocky visitors from outer space. Armed with the knowledge to identify meteorites from ordinary rocks, meteorite hunters will, if successful, collect these space rocks and produce a reasonable income from selling them at star parties, astronomy forums and expos, and to museums if the sample is large enough (Fig. 2.13).

Solar Astronomy—Daytime solar astronomy is a fascinating area that many amateur astronomers pursue. Unlike nighttime astronomy, where the emphasis is on the telescope's ability to act as a light bucket, gathering every faint photon of light

Fig. 2.13 Fragments of the Sutter's Mill meteorite fall collected by NASA (NASA)

possible to produce a viewable image, the Sun is eye damagingly bright and the task becomes limiting the brightness of the Sun to prevent damage to one's eyes.

CAUTION: DO NOT OBSERVE THE SUN WITHOUT PROPER SOLAR FILTERING EQUIPMENT. PLEASE READ THE FOLLOWING:

1. **Never look directly at the Sun with the naked eye or with a telescope, unless the proper solar filter is being used.** Permanent and irreversible eye damage will result without proper protection.
2. **Never use the telescope to project an image of the Sun onto any surface.** Internal heat buildup can damage the telescope and any accessories attached to it.
3. **Never use an eyepiece solar filter on a telescope.** Internal heat buildup inside the telescope can cause these devices to crack or break, allowing unfiltered sunlight to pass through to the eye and cause irreparable damage and blindness.
4. **Never leave the telescope unattended when viewing the Sun.** People and children unfamiliar with the dangers of viewing the unfiltered Sun may do something stupid if left alone with the telescope. Never underestimate the dumbness and stupidity of the general public.

Observing the Sun is a very dynamic activity. Unlike deep sky objects, such as galaxies and nebulae that never change from night-to-night, week-to-week, or year-to-year, the Sun changes from minute-to-minute. Sunspots move across the face of the Sun. With a Hydrogen-Alpha filter or telescope, an observer can watch as solar prominences develop, grow and expand, and fade. The Sun is easily located, and the observer doesn't lose sleep at night. Solar observing does demand a different suite of equipment, including white-light solar filters, H-alpha filters, and dedicated H-alpha telescopes. Unfortunately, solar filtering equipment is solely dedicated for daytime Sun observations, with no applications for nighttime activity (Figs. 2.14 and 2.15).

Fig. 2.14 A Coronado PST Hydrogen-Alpha telescope (Hands-on-Optics archive)

Fig. 2.15 Typical view through a Hydrogen-Alpha telescope (Hands-on-Optics archive)

Chapter 3

Keeping Healthy, Active, and Backyard Astronomy

The Nitrogen in our DNA,
the Calcium in our teeth,
the Iron in our blood,
the Carbon in our apple pies,
were made in the interiors of collapsing stars.
We are made of Star Stuff

—Carl Sagan

The keys to good health in the older years is to maintain good physical health, keep mentally active, eat a good balanced diet, and practice good healthful habits. Whether you are an amateur astronomer, or just a normal civilian mortal (haha!), a balance of all these factors is important to the quality of life after the age of 50.

Good physical health and good eye health are essential for amateur astronomers. Good eye health is an obvious necessity, but to enjoy the hobby, good overall health is also a requirement. Extensive credible information exists on staying healthy as one grows older. There are libraries, websites, and bookstores full of publications and information on health and the aging process.

Genetics plays a large role in the aging process. So there is an element of predestination during aging. But there is plenty that the aging astronomer can control that will extend the quality of life in the later years.

J.L. Chen, *Astronomy for Older Eyes*, The Patrick Moore Practical Astronomy Series, DOI 10.1007/978-3-319-52413-9_3

General Principles of Aging and Health

To summarize from the wealth of information available, here are the general principles of health for growing older:

1. Healthy aging means continually reinventing oneself while passing through landmark ages such as 60, 70, 80, and beyond. It means finding new things to enjoy, learning to adapt to change, staying physically and socially active, and maintaining connection with community and loved ones. Amateur astronomy is an activity that fills these needs in the later years.
2. Experiencing meaningful life and joy does not and should not diminish as one grows older. Amateur astronomy is an activity that can bring meaning to life and be enjoyable in the later years.
3. A person's network of friends changes in the later years, and this poses a great challenge. Staying connected isn't easy as one grows older, even for those who have always had an active social life. Career changes, retirement, illness, moving out of the local area, and death can take away close friends and family members. The older one gets, the more people that are inevitably lost. Fortunately, the astronomy community is quite large, diverse, and easily accessible. Every region of the country has a local astronomy club that offers like-minded people the opportunity to socialize, exchange thoughts and ideas, and share in the experience of discovery. Annual and periodic star parties abound, enabling a senior astronomer to intermingle and converse with others. The opportunities for staying connected are better than ever, especially with the development and acceptance of Internet social networks from Facebook, Twitter, and the like.
4. Exercise to stay healthy as one ages. A 2012 Swedish study found that exercise is the number one contributor to longevity, adding extra years to your life—even if you don't start exercising until the senior years. But it's not just about adding years to life, it's about adding quality of life to those years. Exercise helps maintain strength and agility, increases vitality, improves sleep, gives mental health a boost, and can even help diminish chronic pain. Exercise can also have a profound effect on the brain, helping to prevent memory loss, cognitive decline, and dementia (Figs. 3.1 and 3.2).
5. Protection from the ravages of the environment. The use of sunscreen and wearing protective clothing is recommended while out in the sun. Wearing earplugs when mowing the lawn, watching fireworks, or attending a rock concert will help preserve hearing. Getting older means watching out for the general health of one's body. Life is more enjoyable when in possession of all faculties.

Regular exercise will help maintain physical and mental health, and improve self-confidence and overall outlook on life (Figs. 3.3 and 3.4).

1. Check with a doctor before starting any exercise program. Find out if any health conditions or medications affect the type of exercise chosen.
2. Find an activity that's attractive, fun, and that motivates continuing participation. Some may want to exercise in a group, like in a team sport or class, or some

Fig. 3.1 Wellness and fitness facility (James Chen)

Fig. 3.2 Cardio fitness machines (James Chen)

Fig. 3.3 Lap swimming pool (James Chen)

Fig. 3.4 Weight training at the fitness center (James Chen)

Fig. 3.5 Spin cycling (James Chen)

prefer a more individual exercise like swimming. Just make sure to keep moving. Many gyms and fitness facilities have qualified and certified trainers and physical therapists whose job is to help (Fig. 3.5).

3. Start any physical training slowly. Those new to exercise can begin with a few minutes a day as a good start towards building a healthy habit. Slowly increase the time and intensity to avoid injury.
4. Walking is a wonderful way to start exercising. Exercise doesn't have to mean strenuous activity or a long time at the gym. In fact, walking is one of the best ways to stay fit. Best of all, it doesn't require any equipment or experience and can be done anywhere.
5. Exercise with a friend or family member. Mutual support can help to keep each other motivated with benefits beyond the physical activity. The social contact is beneficial as well.
6. Make yoga sessions a part of the fitness routine. Seniors often experience balance problems as they age. Performing specific yoga routines can help alleviate some balance issues. When thinking about balance, yoga positions that are often envisioned are holding a picturesque half-moon or tree pose. Falls from losing one's balance is a major problem in old age. Falls often occur when moving, transitioning, or adjusting one's position. The key to building balance, coordination, and preventing injury from falling when growing older is in mastering transitions and developing power. Here again, many gyms and fitness facilities have

certified trainers and physical therapists whose job is to conduct yoga classes and help people with balance issues.

7. Plan out a weekly routine for exercise, either self-developed or with the help of a trainer. Be sure to include aerobic exercise and strength training. Include days off to allow the body to rebuild and replenish.

The personal relationship with food will change along with changes to the body as the years progress. Decreased metabolism, changes in taste and smell, and slower digestion may affect the appetite, the foods that can be eaten, and how the body processes food. The key is to figure out how to adapt to changing needs. Healthy eating is important to maintain energy and health.

1. Load up on high-fiber fruits, vegetables, and whole grains. The whole digestive system does slow during aging, so fiber is very important. Consume fiber-rich foods such as whole grains, fruit, and vegetables. The complex carbohydrates that these foods represent will provide more energy over a longer time (Fig. 3.6).
2. Put some effort into making meals look and taste good. The older taste buds may not be as strong and the older adult appetite may not be the same, but the nutritional needs are just as important as ever. If eating isn't as enjoyable as it used to be, there is help. Consult cookbooks, watch the Cooking Channel or Food Network for ways to improve the flavor, the preparation, and the presentation of breakfast, lunch, or dinner (Fig. 3.7).
3. It is important to stay hydrated. Because of physical changes, older adults are more prone to dehydration. Staying hydrated is especially important for those on

Fig. 3.6 Fruits and vegetables (Adam Chen)

Fig. 3.7 The author's favorite—Dim Sum (James Chen)

medications that act as diuretics. So drink plenty of liquids, even when not thirsty. Not getting enough water results in not being as sharp and energetic.

4. Make meals a social event. It's more enjoyable to eat with others than alone. Live alone? Invite other people over. It's a great way to stay in touch with friends and share cooking and cleanup duties.

It is important that older people get enough sleep at night. This seems counter in a hobby that is practiced mostly at night. Sleep is important.

The Mayo Clinic recommends 7 to 9 hours of sleep for those over 50 years. Older adults tend to sleep more lightly and for shorter periods of time than twenty-something adults. There tends to be a number of reasons why older adults fail to get a good night's sleep, from insomnia, sleep apnea, restless legs disorder, periodic limb movement disorder, and many more. Consult a physician for treatment and management of sleep problems.

Weaker bladders and frequent nocturnal bathroom visits often sabotage nighttime sleep patterns for 50+ year olds. This is the main reason the phrase "Grandpa naps a lot!" is heard in many family circles. This is one of the most commonplace problems that affect sleep.

To combat "Midnight bathroom" syndrome, watch fluid intake in the evenings and don't consume caffeinated drinks in the evening. Dry mouth can be remedied by eating a piece of fruit or sucking on an ice cube.

If sleep is still lacking, consult a health professional.

Nighttime astronomy will have an effect on getting enough sleep. The temptation is to let good weather dictate the amount of observing. At a younger age, pulling all-nighters was easy. Getting older means all-nighters may be a thing of the past. A little planning of an observing schedule can alleviate any potential sleep problems. Plan observing nights for the week, and avoid consecutive late night sessions. A lot will depend on the weather cooperating, of course.

Poor sleep habits are often the main causes of low-quality sleep in older adults.

1. Naturally boost the body's melatonin levels at night. Artificial lights at night can suppress the body's production of melatonin, the hormone that makes a person feel sleepy. Use low-wattage bulbs where safe to do so, and turn off the TV and computer at least 1 hour before bed.
2. Make sure the bedroom is quiet, dark, and cool, and the bed is comfortable. Noise, light, and heat can interfere with sleep. Try using an eye mask to help block out light.
3. Develop bedtime rituals. A soothing ritual, like taking a bath or playing music, will help the process of winding down.
4. Go to bed earlier. Adjust the bedtime schedule to match the feeling of being tired, even if that's earlier than it used to be.
5. Increase daytime activities. If too sedentary, the feeling of sleepiness never occurs, or feeling sleepy will occur all of the time. Regular aerobic exercise during the day, at least 3 h before bedtime, can promote good sleep.

There are many good reasons for keeping the brain as active as the body. It is possible to prevent cognitive decline and memory problems by physically exercising, maintaining an active brain and being creative. Don't veg out! It is beneficial to have increased activities and socializing, which increases the use of and will sharpen the brain. This is especially true during retirement where there are no longer job and career challenges. For those new to astronomy, notice how astronomy answers the call for all of the following challenges:

1. Try variations on familiar activities. For some people, it might be games. Other people may enjoy puzzles, learning to play a musical instrument, or trying out

new cooking recipes. Find something that is enjoyed and continue to try new variations and challenges. Those who like crosswords, move to a more challenging crossword series or try a new word game. Learn to play a guitar, piano, or some kind of brass or woodwind instrument. Learn to play whole songs by memory. For those who like to cook, try a completely different type of food, or try baking.
2. Work something new in each day. Don't be limited to elaborate crosswords or puzzles to keep the memory sharp. Try something new and different each day, such as taking a different route to the grocery store or brushing your teeth with a different hand.
3. Explore a completely new subject. Diving into a new subject is a great way to continue to learn. Learn a different language. Learn a new computer language or operating system. Write a book. Learn to play golf. There are many inexpensive classes at community centers or community colleges that allow you to tackle new subjects. Volunteering is also a great way to learn about a new area. Taking classes and volunteering is a great way to boost social connections, which is another brain strengthener. Be a docent at a museum.

Social Activity in the Later Years

There are three competing theories concerning aging and the role that interacting socially plays in self-worth, pleasure, happiness, and longevity.

The activity theory of aging proposes that older adults are happiest when they stay active and maintain social interactions. These activities, such as amateur astronomy, help the elderly to replace lost life roles after retirement and, therefore, resist the social pressures that limit an older person's world. The theory assumes a positive relationship between activity and life satisfaction. Activity theory reflects the functionalist activities that the equilibrium between social activities and happiness that an individual develops in middle age should be maintained in later years. The theory predicts that older adults that face role loss will substitute former roles with other alternatives.

The disengagement model suggests that it is natural for the elderly to disengage from society as they realize that they are ever nearer to death. Since the primary role of individuals is to work or raise families, the elderly will face internal conflicts after retirement when they are separated from these roles. Disengagement, under this theory, allows the elderly to more easily assume different roles.

The continuity theory of normal aging states that older adults will usually maintain the same activities, behaviors, and relationships as they did in their earlier years of life. According to this theory, older adults try to maintain this continuity of lifestyle by adapting strategies that are connected to their past experiences. The continuity theory is considered by some as a subset of the activity theory.

Five decades of gerontological research suggests that the activity model, and to a lesser extent the continuity theory, is more accurate than the disengagement model. Not only is activity beneficial for the community, but it engages older adults both physically and mentally and allows them to socialize with others. This increases feelings of self-worth and pleasure, which are important for happiness and longevity.

Activities represented by astronomy clubs and related public outreach programs enable seniors to stay engaged socially, interacting with people while maintaining and learning new skills and knowledge.

Maintaining Mental Health

Growing older does not mean that mental abilities will necessarily be reduced. There's a lot that can be done to keep the mind sharp and alert. Researchers believe that many of the supposed age-related changes that affect the mind, such as memory loss, are actually lifestyle related. Just as muscles get flabby from sitting around and doing nothing, so does the brain.

A marked decline in mental abilities may be due to factors like prescription medications or disease. Older people are more likely to take a range of medications for chronic conditions than younger people. In some cases, a drug (or a combination of drugs) can affect mental abilities.

The underlying cause of declining mental abilities can be related to certain diseases, such as Alzheimer's disease or dementia. It is worth checking with a doctor to make sure any cognitive changes, such as memory loss, aren't associated with these serious illnesses.

Some of the normal age-related changes to the brain include:

1. Fat and other deposits accumulate within brain cells (neurons), which hinders their functioning.
2. Neurons that die from "old age" are not replaced.
3. Loss of neurons means the brain gets smaller with age.
4. Messages between neurons are sent at a slower speed.

The brain can adapt as the aging process continues. A brain that gets smaller and lighter with age can still function as effectively as a younger brain. For example, an older brain can create new connections between neurons if given the opportunity. There is evidence to suggest that mental abilities are "shared" by various parts of the brain so, as some neurons die, their roles are taken up by other neurons. One study, offered on the PBS special series *The Brain,* examined a group of convent nuns over a prolonged period. The nuns practiced intense devoutness, Bible study, and service, and there was no evidence of mental health issues such as dementia or Alzheimer's. Autopsies following their passing reveal physical signs of Alzheimer's, but in life there was no mental deficiencies, leading researchers to conclude their brains had "rewired" around the damaged areas of the brain.

As previously mentioned, maintaining physical fitness, having an active brain, eating a healthy diet, and staying social are important factors in keeping good mental health as you get older. Regular exercise can improve the brain's memory, reasoning abilities, and reaction times. A good diet avoids the complexities of obesity. Glucose is the brain's sole energy source, so eat a balanced diet and avoid extreme low carbohydrate diets. The combination of low-carb diets and aging is not a good combination.

One of the challenges as we grow older is maintaining a good and healthy memory. Good recall is a learned skill. There are ways to improve a failing memory no matter what age.

Suggestions include:

1. Have good situational awareness. Pay attention to whatever it is that needs to be remembered. For example, take note of setting down house and car keys when other things are demanding attention.
2. Use memory triggers, like association or visualization techniques. For example, link a name you want to remember with a mental picture.
3. Practice using your memory. For example, try to remember short lists, such as a grocery list. Use memory triggers to help you "jump" from one item to the next. One type of memory trigger is a walking route that you know well. Mentally attach each item on your list to a landmark along the route. For example, imagine putting the bread at the letter box, the apples at the next-door neighbor's house, and the meat at the bus stop. To remember the list, you just have to "walk" the route in your mind.

Getting older doesn't necessarily mean that the mind stops working as well as it once did. However, some of the conditions and events more common to older age that affect brain function include:

1. Atherosclerosis
2. Dehydration
3. Dementia
4. Alzheimer's disease
5. Depression
6. Diabetes mellitus
7. Heart disease
8. Medications—Often, medications can lead to chronic fatigue. Prescribed medicines should be regularly reviewed with a doctor so that unwanted side effects are avoided, and drugs should be discontinued if they are no longer required.
9. Poor nutrition, vitamin deficiency
10. Parkinson's disease
11. Stroke

Many conditions can be managed. Consult with a physician, or if necessary a neurologist for lifestyle strategies. Many of the conditions that may affect brain function can be managed effectively. The following factors have all proved to be important:

1. Lifestyle and diet changes
2. Monitoring tests for hypertension, cholesterol, and diabetes
3. Medications

The bottom line for good mental health in the retirement years are:

1. Researchers believe that many of the supposed age-related changes which affect the mind, such as memory loss, are actually lifestyle related.
2. Keeping an active body is crucial for an active mind.
3. Recognize some of the conditions and events more common to old age that may hinder brain function including dementia, Parkinson's disease, and atherosclerosis.

More on Exercise

Being physically active on a regular basis is one of the healthiest things you can do for yourself. Studies have shown that exercise provides many health benefits and that older adults can gain a lot by staying physically active. Even moderate exercise and physical activity can improve the health of people who are frail or who have diseases that accompany aging.

Physical fitness is important for good mental health. Some conditions that can affect the brain's ability to function, such as stroke, are associated with diet, obesity, and sedentary lifestyle choices. Keeping an active body is crucial if you want an active mind. Suggestions include:

1. At least 30 min of moderate exercise every day delivers an oxygen boost to the brain.
2. Exercising in three 10-min blocks is enough to deliver significant health benefits.
3. Regular exercise can improve the brain's memory, reasoning abilities, and reaction times.
4. Avoid the complications of obesity (such as diabetes and heart disease) by maintaining a healthy weight for your height.
5. Don't smoke and avoid drinking to excess.

Being physically active can also help you stay strong and fit enough to keep doing the things you like to do as you get older. The act of carrying a telescope and mount should be easy, not a chore. Making exercise and physical activity a regular part of your life can improve your health and the normal activities of astronomy, besides sitting and observing, of sitting up for a night's observing less stressful.

Be as active as possible. Regular physical activity and exercise are important to the physical and mental health of almost everyone, including older adults. Staying physically active and exercising regularly can produce long-term health benefits and even improve health for some older people who already have diseases and disabilities. That's why health experts say that older adults should aim to be as active as possible.

Although exercise and physical activity are among the healthiest things you can do for yourself, some older adults are reluctant to exercise. Some are afraid that exercise will be too hard or that physical activity will harm them. Others might think they have to join a gym or have special equipment. Yet, studies show that "taking it easy" is risky. For the most part, when older people lose their ability to do things on their own. It doesn't happen just because they've aged. It's usually because they're not active. Lack of physical activity also can lead to more visits to the doctor, more hospitalizations, and more use of medicines for a variety of illnesses.

Scientists have found that staying physically active and exercising regularly can help prevent or delay many diseases and disabilities. In some cases, exercise is an effective treatment for many chronic conditions. For example, studies show that people with arthritis, heart disease, or diabetes benefit from regular exercise. Exercise also helps people with high blood pressure, balance problems, or difficulty walking. Balance problems and challenges to walking are major problems of old age. Remember the "I've fallen and I can't get up" ads on TV?

Regular exercise will prevent dropping a telescope on the deck or patio (remember the introduction of this book?). Resistance training in the form of weight lifting and resistance machine workouts have prevented any further incidents of this kind for this author.

Regular, moderate physical activity can help manage stress and improve your mood. Being active on a regular basis may help reduce feelings of depression. Studies also suggest that exercise can improve or maintain some aspects of cognitive function, such as your ability to shift quickly between tasks, plan an activity, and ignore irrelevant information.

Some people may wonder what the difference is between physical activity and exercise. Physical activities are activities that get your body moving such as gardening, walking the dog, and taking the stairs instead of the elevator. Exercise is a form of physical activity that is specifically planned, structured, and repetitive such as weight training, spin cycling, or an aerobics class. Including both in your life will provide you with health benefits that can help you feel better and enjoy life more as you age.

Stress Is Destructive to the Body

Avoid stress. Stress is any change in the environment that requires your body to react and adjust in response. Stress is an automatic response developed evolutionarily in the human body as a way to protect it from predators and other threats. Faced with danger, the body kicks into high gear, flooding the body with hormones that elevate your heart rate, increase your blood pressure, boost your energy, and prepare it to deal with the problem. The body reacts to these changes with physical, mental, and emotional responses (Fig. 3.8).

Stress is a normal part of life. Many events that happen to you and around you—and many things that you do yourself—put stress on your body. You can experience good or bad forms of stress from your environment, your body, and your thoughts.

Fig. 3.8 The epitome of no stress (James Chen)

Backyard astronomy is a good form of stress relief. For those newly retired, and those long since retired, hopefully the stress from earning a living has been greatly reduced. No longer is there a boss breathing down your neck!

However, stress can still take many forms and still be damaging even if you're retired. Studies have shown that these sudden emotional stresses, especially losing your temper and anger, can trigger heart attacks, arrhythmias, and even sudden death. Although this happens mostly in people who already have heart disease, some people don't know they have a problem until acute stress causes a heart attack or something worse.

Consider the following:

1. Forty-three percent of all adults suffer adverse health effects from stress.
2. Seventy-five percent to 90% of all doctor's office visits are for stress-related ailments and complaints.
3. Stress can play a part in problems such as headaches, high blood pressure, cardiac problems, diabetes, skin conditions, asthma, arthritis, depression, anxiety, and weight issues associated with overeating.
4. The Occupational Safety and Health Administration (OSHA) declared stress a hazard of the workplace. Stress costs American industry more than $300 billion annually.
5. The lifetime prevalence of an emotional disorder is more than 50%, often due to chronic, untreated stress reactions.

The American Psychological Association recommends the following actions to reduce stress:

1. Identify what's causing stress—Monitor your state of mind throughout the day. If you feel stressed, write down the cause, your thoughts, and your mood. Once you know what's bothering you, develop a plan for addressing it. That might mean setting more reasonable expectations for yourself and others or asking for help with household responsibilities, job assignments, or other tasks. List all your commitments, assess your priorities, and then eliminate any tasks that are not absolutely essential.
2. Build strong relationships—Relationships can be a source of stress. Research has found that negative, hostile reactions with your spouse cause immediate changes in stress-sensitive hormones, for example. But relationships can also serve as stress buffers. Reach out to your spouse, other family members, or close friends and let them know you're having a tough time. They may be able to offer practical assistance and support, useful ideas or just a fresh perspective as you begin to tackle whatever's causing your stress.
3. Walk away from anger—Before reacting, take time to regroup by counting to ten. Then reconsider. Walking or other physical activities can also help work off steam. Plus, exercise increases the production of endorphins, the body's natural mood-booster. Commit to a daily walk or other form of exercise—a small step that can make a big difference in reducing stress levels.
4. Rest your mind—Stress keeps more than 40% of adults lying awake at night. To help ensure getting the recommended 7 to 8 hours of shut-eye, cut back on caffeine, remove distractions such as television or computers from the bedroom, and go to bed at the same time each night. Research shows that activities like yoga and relaxation exercises not only help reduce stress, but also boost immune functioning.
5. Get help—Still feel overwhelmed, consult with a psychologist, or other licensed mental health professional who can help you learn how to manage stress effectively. He or she can help identify situations or behaviors that contribute to chronic stress and then develop an action plan for changing them.

By keeping healthy and active, anyone can continue with (or for beginners, start) their love affair with backyard astronomy.

Chapter 4

Older Eyes, Cataracts, Lasik and Laser Eye Surgery

If a little kid ever asks you just why the sky is blue, you just look at him or her right in the eye and say, "It's because of quantum effects involving Rayleigh scattering combined with the lack of violet photon receptors in our retinae."

—Phillip C. Plait

Human eyes undergo changes as we age. Good diet and avoiding extreme bright light may have a lessening impact on eye health, but in many cases there is still a deterioration of eyesight. Genetics will play a large role in the aging process and how well sight is preserved as one ages. So there is an element of predestination during aging. But there is plenty of room for changes that the aging astronomer can make that will extend good eyesight and, in turn, the quality of life in the later years. As an aging astronomer, there are strategies that will keep observing time stretching well into later years.

The key to maintaining good eye health is an annual visit to the ophthalmologist. The eye doctor can monitor the status of the eyes, diagnose and assess any changes, and either through prescriptions, eye drops, or as a last resort, perform surgery to correct the various malfunctions of the eyes as we age. Many eye problems can be prevented or corrected if detected in their early stages. An ophthalmologist is a medical doctor, either an M.D. or D.O., with special training to diagnose or treat eye disorders.

Eye health is not to be treated lightly. The Internet is informative, but will not diagnose or provide individualized treatment for any eye problems. The American Academy of Ophthalmology recommends a comprehensive eye examination every

J.L. Chen, *Astronomy for Older Eyes*, The Patrick Moore Practical Astronomy Series, DOI 10.1007/978-3-319-52413-9_4

2–4 years for people between the ages of 40–64 years, with yearly checkups for those over 65 years old. Do yourself a favor. See an ophthalmologist once a year after your 50th birthday. Especially when your hobby is astronomy and your eyes are your primary tools.

Nutrition and Good Eye Health

Before this chapter delves into the disorders of the eye, the question needs to be asked: Are you eating the foods that are best for your eyes? There's more to eye nutrition than Bugs Bunny's diet of just eating carrots. The following foods are recommended by doctors, dietitians, and health authorities for both eye health and overall health, and help protect against sight-threatening diseases:

1. Fish—Cold-water fish such as salmon, tuna, sardines, and mackerel are rich in omega-3 fatty acids which may help protect against dry eyes, macular degeneration, and even cataracts. Omega-3 fatty acid is also a "good" HDL recommended for overall health by keeping cholesterol and the "bad" LDL levels in check.
2. Leafy greens—Spinach, kale, Swiss chard, and collard greens, to name a few, are full of lutein and zeaxanthin, plant pigments that some health experts claim can help stem the development of macular degeneration and cataracts. Broccoli, peas, and avocados are also good sources of the lutein and zeaxanthin antioxidant duo. Again, consumption of leafy greens is recommended for good overall general health.
3. Eggs—The vitamins and nutrients in eggs, including lutein and vitamin A, promote eye health and function. Consult your physician on consuming eggs, as there are concerns over their role in cholesterol development. And there are some, including this author, who are allergic to egg whites, which preclude their consuming them.
4. Whole grains—A diet containing foods with a low glycemic index (GI) can help reduce your risk for age-related macular degeneration. Spelt, buckwheat, quinoa, rye, wild rice, barley, and bulgur are all recommended as part of a diet good for eyes and the rest of the body.
5. Citrus fruits and berries—Oranges, grapefruits, lemons, and berries are high in vitamin C, which may reduce the risk of cataracts and macular degeneration. The benefits of vitamin C have historically been recognized for maintaining a healthy immune system. These fruits are best eaten raw and uncooked, as vitamin C is easily broken down by heat.
6. Nuts and seeds—Pecans, pistachios, walnuts, almonds, brazil nuts, filberts, and others are rich in omega-3 fatty acids and vitamin E that boost eye health. Brazil nuts are also a great source for zinc in the diet. Sunflower seeds are an excellent source for vitamin E and zinc.
7. Deeply colored vegetables and fruits—Finally, the carrot gets its due. Carrots, along with tomatoes, bell peppers, strawberries, pumpkin, corn, and cantaloupe, are all excellent sources of vitamins A and C. The compounds called carotenoids

that give these fruits and vegetables their yellow, orange, and red pigments are also contained in these vegetables, and are thought to help decrease the risk of many eye diseases. Bugs Bunny is finally happy!

8. Legumes—Kidney beans, lima beans, black-eyed peas, and lentils are good sources of bioflavonoids and zinc and can help protect the retina and lower the risk for developing macular degeneration and cataracts. For general good health, beans and legumes are a great source for fiber in the diet.
9. Beef—Apologies to those vegetarians and vegans out there. In moderation, lean beef in the diet can boost eye health. Beef contains zinc, which helps the human body absorb vitamin A and may play a role in reducing risk of advanced age-related macular degeneration.

Disorders of the Aging Eye

No matter the diet or good health practices exercised, the eye structure experiences these normal changes during the aging process:

1. Reduced pupil size—People in their 60s need three times more ambient light for comfortable reading than those in their 20s. The iris at age 20 can open to a diameter of 7 mm. By age 60, the iris opening is reduced to approximately 5.5 mm. The muscles controlling pupil size and reaction to light lose their strength, causing the pupil to become smaller and less responsive.
2. Dry eyes—The body produces less tears as one ages, necessitating the need for artificial tears and other medications prescribed by the ophthalmologist.
3. Loss of peripheral vision—The size of the eye's field-of-view decreases by 1–3° per decade of life.
4. Decreased color vision—Cells in the retina responsible for normal color vision decline in sensitivity with age. Colors become less bright and contrast between colors is less noticeable. Colors towards the blue end of the spectrum can appear particularly faded.
5. Eye strain—Medically referred to as neurasthenia, it occurs with overuse of the eyes particularly from viewing objects at a fixed distance over a long period of time. It is a common problem in all age groups, especially in this day and age with long hours of computer use, watching television, and reading. As one ages, you are more prone to eye strain as the pupillary reaction to light may be slow or not as effective as in the younger years. In addition, the lens loses its elasticity and tear production may be lesser than normal in the senior years. Although eye strain can be a source of significant discomfort and even cause other symptoms like headaches, it is largely reversible and simple lifestyle changes can prevent it. For amateur astronomers, the one-eyepiece style of observing through a telescope can result in eye strain. The answer is to use binoviewers. See the next chapter.

The following are the medical eye problems that face all people, astronomers, or civilians, that can be diagnosed and treated by ophthalmologists. These are present in order of least troublesome and treatable to major issues that potentially lead to the loss of sight. This information was gathered from the American Academy of Ophthalmology:

Presbyopia—From birth to about age 40, the lens of the eye is flexible and changes shape easily, allowing people to focus their eyes close-in and also far away. The lens in our eye becomes less flexible as we get older. By the late 30s or early 40s, reading glasses or bifocals becomes necessary as the lens becomes less flexible and cannot change shape for close-in focusing. Other treatments include mono-vision LASIK surgery and kamra corneal inlay. The simple remedy for presbyopia is the use of reading glasses, or bifocals for eyeglass wearers.

Floaters—Close your eyes. See those specks or small clouds moving about your vision. Those are floaters. The gel-like vitreous inside the eye begins to liquify with age and pull away from the retina. Floaters are tiny clumps of gel inside the vitreous that fills the inside of the eye. The specks cast a shadow on the retina, which converts the light and shadows to signals to the brain and registers as floaters. An increase in floaters indicates a change within the vitreous gel. A small minority of patients with new floaters will develop a retinal tear or detachment. Only by examining the retina can an ophthalmologist determine if someone has a tear or detachment. The ophthalmologist is the only person that can treat this disorder.

Cataracts—Cataracts are clouding of the lens in the eye, causing the eyesight to become blurred. Cataracts cause blurred vision, poor night vision, glare or light sensitivity, double vision in one eye, and fading of colors. A person with cataracts will often require brighter light to read. An option for treatment is cataract surgery, to be discussed in greater detail later in this chapter.

Glaucoma—One of the leading causes of blindness, glaucoma is a disease of the optic nerve, which carries signals from your eyes to your brain. Glaucoma exists when the pressure inside the eye damages the optic nerve. This intraocular pressure is caused by the inability of the clear liquid that flows in the eye to drain improperly. This is a serious condition, and can only be treated by an ophthalmologist. Early detection of glaucoma is important. The available treatments can stop the deterioration of the damage, but cannot reverse the damage. People can get glaucoma because of aging, elevated eye pressure, family history of glaucoma, and past eye injuries. Those of African or Spanish descent have a greater incidence of glaucoma.

Macular Degeneration—The macula is the small central area of the retina responsible for fine detail detection. Macular degeneration is a breakdown of the macula and the main cause is aging and thinning of the tissues of the macula, with a resulting gradual loss of vision. Rapid macular degeneration and resulting loss of vision can occur from abnormal blood vessels leaking fluid or blood under the macula. According to a European study published in the October 2008 issue of *Archives of Ophthalmology*, ultraviolet and blue light, especially when combined with low blood plasma levels of vitamin C and other antioxidants, is associated with the development of macular degeneration. This condition is very disturbing to an astronomer, resulting in the ability to see a wide field of stars, but unable to detect any details. Again, the ophthalmologist has treatments to slow and stabilize the condition, but there is no cure.

Diabetic Eye Problems—Older adults often suffer from either Type 1 or Type 2 diabetes, which is the body's inability to use or store sugar properly. Diabetes can cause changes to the human circulatory system. Diabetes affects vision by causing cataracts, glaucoma, and damage to the eye's blood vessels. When internal blood vessels leak or cause scar tissue to form within the eye, blurring and distortion can result. This condition, called diabetic retinopathy, can be prevented by good control of blood sugar and blood pressure. In extreme cases, an ophthalmologist can perform laser surgery to halt further damage and loss of sight. It is mandatory that those amateur astronomers who have diabetes work with their endocrinologist and control their diabetes with diet and with insulin injections for Type 1, or medication with Type 2 diabetes. People with Type 2 diabetes should be examined at the time of diagnosis. People with type 1 should be seen within 5 years of diagnosis. Pregnancy may accelerate diabetic retinopathy.

Cataracts and the Impact on Telescope Use

With the use of the eyes as the chief instrument of enjoying the hobby, the formation of cataracts is a major concern among astronomers. Cataracts form with age. About half the people in the United States have some degree of cataract formation by the age of 80. Cataracts may also form because of trauma and radiation exposure. A small percentage of infants are born every year with cataract. Occasionally, eye surgery accelerates cataract formation. Prolonged exposure to sunlight, diabetes, and smoking increases the risk of the development of cataracts (Fig. 4.1).

So what are cataracts exactly? Clumps of protein or yellow-brown pigment develop and form in the lens, reducing the transmission of light to the retina at the back of the eye. Cataracts can affect one or both eyes, and develop slowly, such that they go unnoticed in the early stages.

Americans Who Had Cataracts in 2010

Age Group	Percentage With Cataracts
50-54	5.2%
55-59	9.1%
60-64	15.4%
65-69	24.7%
70-74	36.5%
75-79	49.5%
80+	68.3%

Fig. 4.1 Percentage of the US population with cataracts (allaboutvision.com)

How does the development of cataracts affect an amateur astronomer? The brightness of the objects being viewed is obviously diminished. But this can be tricky to detect. A lunar observer could go for years and not notice the dimming of the image when viewing through the eyepiece, due to the brightness of the Moon.

Planetary observers may start noticing diminished detail. The swirls and eddies of Jupiter's cloud belts become difficult to spot. The Cassini Division in the rings of Saturn is not quite as definitive. During Mars oppositions, the already low contrast Martian details become even more difficult to observe as the contrast is lessened by the cataracts. When observing any of the planets, astronomers with developing cataracts may notice a faint glow around the planets, somewhat like a halo, where previously there was just blackness.

Deep sky observers and double star observers with developing cataracts will find the stars in clusters and their favorite double stars don't quite sparkle like they used to. The ability to split closely separated doubles becomes a challenge. Some globular clusters that used to easily resolve become mere fuzzballs. Filaments of nebulosity that surround emission or reflection nebulae may become less visible.

The early stages of cataracts for an amateur astronomer can be countered by switching to either a larger aperture telescope or using a refractor telescope. The larger aperture is effective because it gathers more light to send to the eyepiece. The additional light counteracts somewhat the dimmer and lessened detail caused by the early stages of some types of cataracts. However, there is one type of cataract, a posterior subcapsular cataract, where the vision gets dramatically worse with brighter light.

The use of refractors will take a little further explanation.

A refractor is an unobstructed telescope design, with no mirror centered along the light path. The physics of the light image is of an Airy disk. In its unobstructed light path, 80% of the light energy is in the central lobe of the Airy disk, with the remainder 20% energy spread in the first and second lobes of the Airy Disk (Fig. 4.2).

The effect of central obstruction, such as that caused by the secondary mirror in a Newtonian or Cassegrain telescope, is to transfer more light from the center of the Airy disk to the outer lobe rings. The SCT example is of a 33% central obstruction. This would cause the first ring to become almost four times brighter

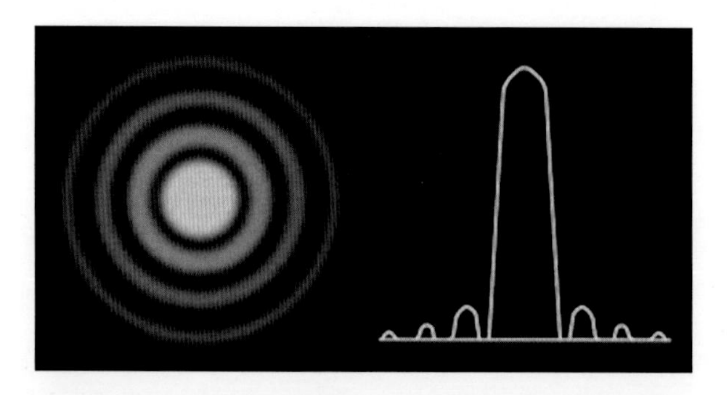

Fig. 4.2 Conceptual depiction of the Airy disk (Hands-on-Optics archive)

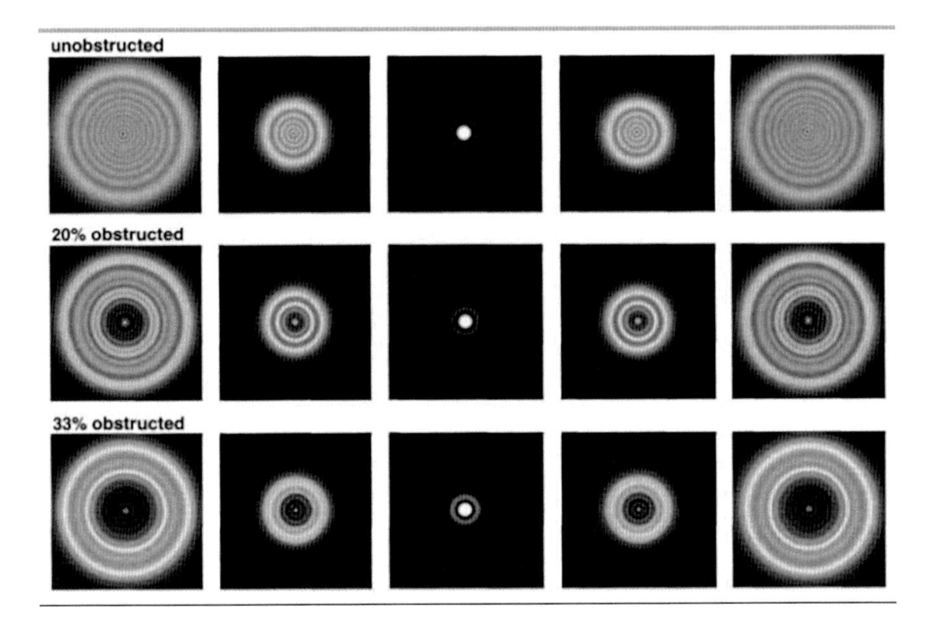

Fig. 4.3 Comparison of a star Airy disk with unobstructed (refractor), 20% obstructed (Newtonian), and 33% obstructed (SCT) optics (Hands-on-Optics archive)

while the central disk would drop in brightness by nearly a factor of 2. The result is blob-like star images that lack the intense tight sparkle that refractor star images have. Planetary details become less distinct with less detail. Overall the image quality is worse when a secondary obstruction is present, with a distinct fall off in contrast.

Thus, a refractor is capable of a much tighter, more intense star image, with nebulae and planets viewed with greater intensity and contrast. The increased contrast available through the use of a refractor can be an effective counter to the effects of the early onset of cataracts and other eye deficiencies of age. This image contrast improvement is even greater using an apochromatic refractor versus an achromatic refractor (Fig. 4.3).

The downside using a refractor is the increased cost of refractor per inch of aperture. Achromats and apochromats are reasonably affordable in apertures up to 100–120 mm. Alt-az and German equatorial mounts for this size of refractor are also quite attainable. However, 130 mm apertures and up, the cost of both telescope and mount increase exponentially. For example, a 200 mm Newtonian on a Dobsonian mount will cost under $400. For example, a 200 mm SCT can be had in the $1000 range. An apochromatic refractor telescope of 200 m aperture with an appropriate mounting can be equivalent in cost to a Mercedes-Benz or BMW car! And is virtually immobile. A 102 mm refractor is the recommended size for a refractor, and is a manageable size for the aging astronomer.

Cataract Surgery

Eventually, the inevitable occurs. Cataract surgery is required. Modern cataract surgery is a common form of eye surgery performed in the United States, and is one of the safest and most effective surgical procedures performed today. More than three million cataract surgeries are performed in the United States every year, with a 98% or better success rate with excellent visual outcomes. That's a great batting average!

While rare, complications after cataract surgery can occur. Most problems can be treated successfully, but there is a risk of some degree of vision loss. Remember, this still is surgery, and surgery has risks. This is best discussed with your eye surgeon. The complications include:

1. Inflammation or infection of the eye
2. Bleeding in the eye
3. Swelling of the cornea
4. Detachment of the retina
5. Increased pressure inside the eye
6. Dislocation of the implanted lens
7. Accumulation of fluid in the retina
8. Drooping eyelid

So, what exactly is the cataract surgery procedure. The short, crude explanation is the surgeon makes two small incisions in the eye and removes the cataract with an ultrasound probe. An artificial lens is inserted into the eye using the same incision. Zip-zap zowie, new fresh vision in place of the old fuzzy cloudy vision (Fig. 4.4).

Cataract surgery is a procedure used to remove the natural lens in the eye when it becomes clouded. The natural lens is replaced with a plastic, artificial lens that is permanent, requires no care, and can significantly improve vision. Newer artificial lenses can have the natural focusing ability of a young lens, allowing for distance and some near vision, as well. This is a major benefit for those lifelong eyeglass wearers, with the post-procedure result of wearing either a weaker prescription set of eyeglasses or none at all!

Cataract removal is one of the most frequently and commonly performed surgical procedures in the world. The surgery is typically an outpatient procedure that takes less than an hour. Most patients are awake during the procedure and need only local anesthesia. If you need to have cataracts in both eyes removed, you will typically have two separate surgeries. This way, the first eye can heal before the second eye surgery.

There are two types of cataract surgery:

1. Small incision cataract surgery (as previously and crudely described) involves making a small incision in the side of the cornea (the clear outer covering of the eye) and inserting a tiny probe into the eye. The probe emits ultrasound waves that soften and break up the lens into little pieces so it can be suctioned out. This process is called phacoemulsification.

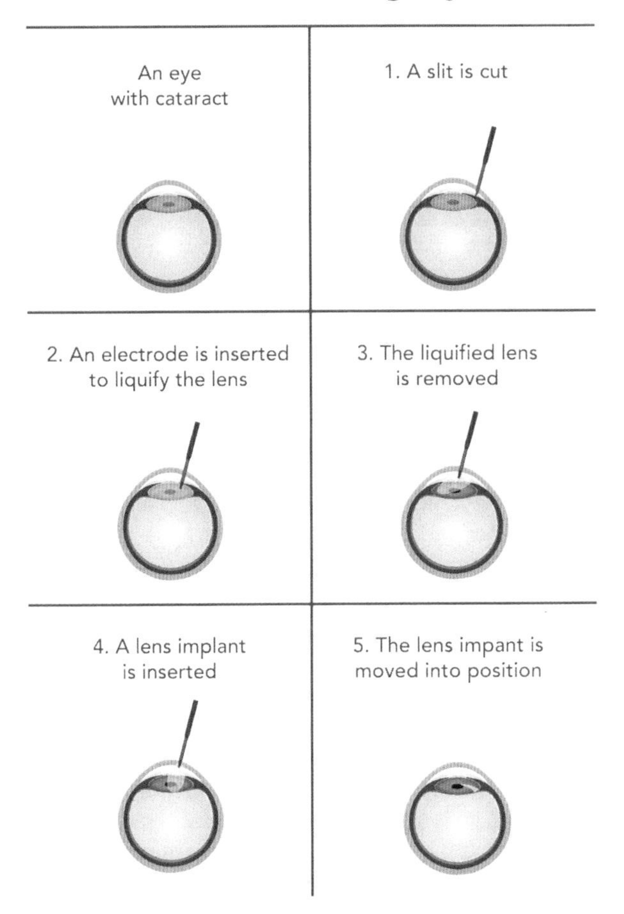

Fig. 4.4 Graphical depiction of cataract surgery (Adam Chen)

During this procedure, the surgeon removes the cataract but leaves most of the thin outer membrane of the lens, called the lens capsule, in place. The incision made for this procedure is so small that the surgeon generally does not need to use sutures to close the opening.

2. Extracapsular surgery requires a somewhat larger incision in the cornea to allow the lens core to be removed in one piece. This approach may be used if the cataract has advanced to the point where phacoemulsification can't break up the clouded lens. If yearly periodic eye examinations have been a practice, it is unlikely the cataracts will have reached this stage resulting in this slightly more complicated procedure. Through this incision, your surgeon opens the lens capsule, removes the central portion of the lens, and leaves the capsule in place.

Once the eye surgeon removes the natural lens, he or she generally replaces it with a clear plastic lens called an intraocular lens (IOL). The IOL is placed in the lens capsule that was left in the eye. The artificial lens can focus light onto the back of the eye and improve vision.

Prior to the procedure, make sure to talk with the ophthalmologist about your astronomy requirements, particularly the need to have a wrinkle-free insertion of the IOL.

When implanting an IOL is not possible because of other eye problems, contact lenses and, in some cases, eyeglasses may be able to correct vision.

Intraocular lenses come in three basic forms: monofocal, astigmatic (toric), and multifocal lenses.

Monofocal lenses are the most commonly implanted lenses. They have the same power in all areas of the lens. They can have a fixed focus or allow for changes in focus.

Fixed focus monofocal IOLs can provide excellent distance vision. However, since these lenses have a fixed focus set for distance vision, you may need to use reading glasses for good near vision. These are the recommended lenses for amateur astronomers. These monofocal lenses allow for the greatest sharpness and contrast required for observing through a telescope.

Accommodating monofocal IOLs is a relatively new lens option that can be used for patients who want both good distance and near vision without the use of eyeglasses or contact lenses. These lenses also have a single focusing power. However, they can shift from focusing on distance objects to focusing on near ones by physically moving inside the eye in response to the focusing action of the eye muscles. This type of lens has a downside. Many patients have noted glare and halos around lights, decreased sharpness of vision, and lack of contrast, especially at night or in dim lighting conditions. Not recommended for amateur astronomers.

Astigmatic (toric) IOLs are monofocal IOLs that have astigmatism correction in them. They can be used for patients who suffer from high astigmatism and want to reduce it.

Multifocal lenses are like bifocal eyeglasses. Several areas of the lens have different powers, which allow individuals to see clearly at far, intermediate and near distances. However, these multifocal lenses are not suitable for everyone. For some individuals, they may cause more problems with night vision and glare than monofocal IOL lenses. Caution is advised for amateur astronomers because of the glare issues and lessened low-light contrast.

Post-Cataract Surgery Recovery and Issues

Since cataract surgery is normally performed as an outpatient procedure, going home on the day of the surgery is allowed so long as someone else is driving. Help is needed at home for a few days because the doctor may limit activities such as bending and lifting.

It is normal to feel itching and some mild discomfort after cataract surgery. There may be temporary fluid discharge from the treated eye and don't be surprised by light sensitivity. Avoid rubbing or pressing on the surgically repaired eye. Try not to bend from the waist to pick up objects on the floor. Do not lift any heavy objects. Movement is limited to walking, climbing stairs, and doing light household chores.

The doctor may prescribe medications to prevent infection and control eye pressure. After a few days, the eye should be comfortable. Normal activities can be resumed within about 8 weeks.

During this healing time, the ophthalmologist will want to monitor eye health and vision. In many cases, refractive testing will occur to check if eyeglasses are still needed for distance and reading activities. Most patients benefit from some form of eyeglasses or contact lenses for optimum vision. Most people need to wear glasses after cataract surgery, at least for some activities. Typical follow-up visits occur 1 day, 1 week, 3–4 weeks, 6–8 weeks, and 6 months after surgery.

Following cataract surgery, regular eye exams are needed to monitor eye health and vision. For amateur astronomers, have your eyes checked at least once a year.

There is a postoperative condition called "secondary cataract" or "aftercataract." This occurs when the lens capsule, the membrane that wasn't removed during surgery and supports the lens implant, becomes cloudy and impairs vision. Another term for this condition is posterior capsular opacification (PCO) (Fig. 4.5).

A secondary cataract can develop months or years after cataract surgery. Cataract symptoms return and vision becomes blurry again. Cell growth on the back of the capsule gradually clouds eye vision.

There is no way to know who may develop clouding of the lens capsule after cataract surgery. Up to 50% of cataract surgery patients experience this problem (Fig. 4.6).

Treatment for a secondary cataract is fairly simple. It involves a technique called YAG laser capsulotomy, in which a laser beam makes a small opening in the

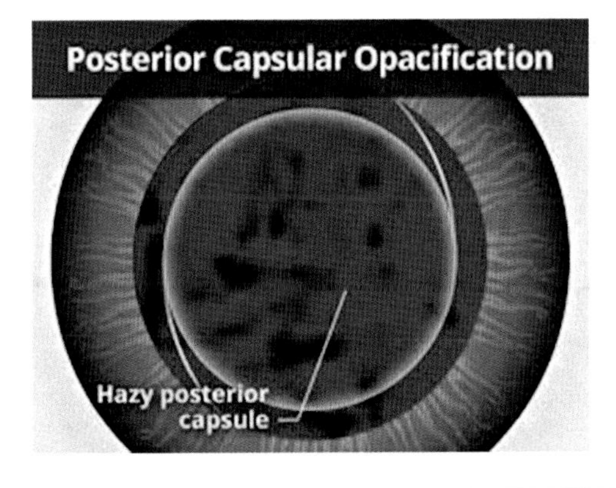

Fig. 4.5 Posterior capsular opacification (PCO) (Allaboutvision.com)

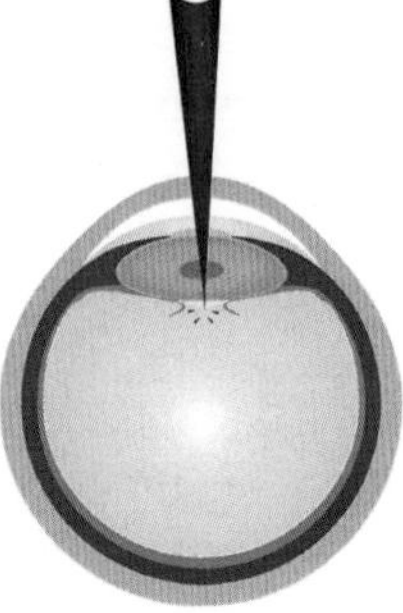

Fig. 4.6 A schematic of YAG laser capsulotomy (Adam Chen)

clouded capsule to allow light through. It is a painless outpatient procedure that usually takes less than 5 min. Most patients are asked to stay in the doctor's office for about an hour to make sure the eye pressure doesn't increase. Most patients immediately experience improved vision, while some experience gradual improvement over several days.

LASIK Surgery

One of the great advancements in eye care in the last 50 years has been the development of techniques for correcting myopia (nearsightedness) and hyperopia (farsightedness).

The most common type of laser vision correction, LASIK is an acronym for Laser-Assisted in situ Keratomileusis, a surgical procedure used to repair nearsightedness, farsightedness, and astigmatism by reshaping the cornea. It is unique in that it also corrects lesser-known problems like "halos," "starbursts," and "ghost images," drastically improving night vision.

LASIK surgery is a controversial procedure among the amateur astronomy community. There is anecdotal evidence that both supports the procedure and those against the procedure. As with any surgery, consultation with the ophthalmologist is required. The discussion should include the fact that astronomy is an important activity. Anecdotal evidence exists that many amateur astronomers who have had LASIK surgery suffer aftereffects that impair their enjoyment of their favorite hobby, with blurriness or halos as a common complaint. Again, remember this is a form of surgery, and surgeries have inherent risks.

For the over-50 years old amateur astronomer, LASIK surgery is not recommended since the likelihood of cataracts development is high. Cataract surgery with an IOL will correct the individual's vision and in many cases is covered by insurance. LASIK surgery is not covered by insurance and is an out-of-pocket expense.

Most corrective measures, including glasses, contacts, and conventional laser surgery, only address issues called lower-order aberrations, providing a satisfactory solution for many people. However, higher-order aberrations, which LASIK targets, still detract from overall quality of vision.

LASIK is a minimally invasive operation during which the patient remains awake, and anesthetic drops are administered to completely numb the eyes. The correction itself takes less than 10 min per eye, with the entire process lasting a few hours. Side effects are generally few.

How is LASIK surgery performed?

1. First, ophthalmologist uses either a mechanical surgical tool called a microkeratome or a femtosecond laser to create a thin, circular "flap" in the cornea.
2. The surgeon folds back the hinged flap to access the underlying cornea and removes some corneal tissue using an excimer laser.
3. This highly specialized laser uses a cool ultraviolet light beam to remove ("ablate") microscopic amounts of tissue from the cornea to reshape it so it more accurately focuses light on the retina for improved vision.
4. For nearsighted people, the goal is to flatten the cornea; with farsighted people, a steeper cornea is desired.

Excimer lasers can also correct astigmatism by smoothing an irregular cornea into a more normal shape. It is a misconception that LASIK cannot treat astigmatism.

After the laser reshapes the cornea, the flap is then laid back in place, covering the area where the corneal tissue was removed. Then the cornea is allowed to heal naturally.

The healing and recovery process is fairly simple, requiring only topical anesthetic drops, and no bandages or stitches are required (Fig. 4.7).

Common LASIK complications and side effects are listed below:

1. Temporary discomfort and vision disturbances—Discomfort during the first few days following LASIK surgery, such as mild irritation and light sensitivity, is normal and to be expected. During the first few weeks or months you also may experience: halos; glare and starbursts in low-light environments, especially at night; dry eye symptoms; hazy vision; and reduced sharpness of vision. In the vast majority of cases, these problems are temporary and disappear completely within 3–6 months.
2. Flap complications—The LASIK procedure involves the creation of a thin hinged flap on the front surface of the cornea. This is lifted during surgery for laser reshaping of the eye. The flap is then replaced to form a natural bandage. If the LASIK flap is not made correctly, it may fail to adhere properly to the eye's surface or microscopic wrinkles called striae could develop in the flap. These

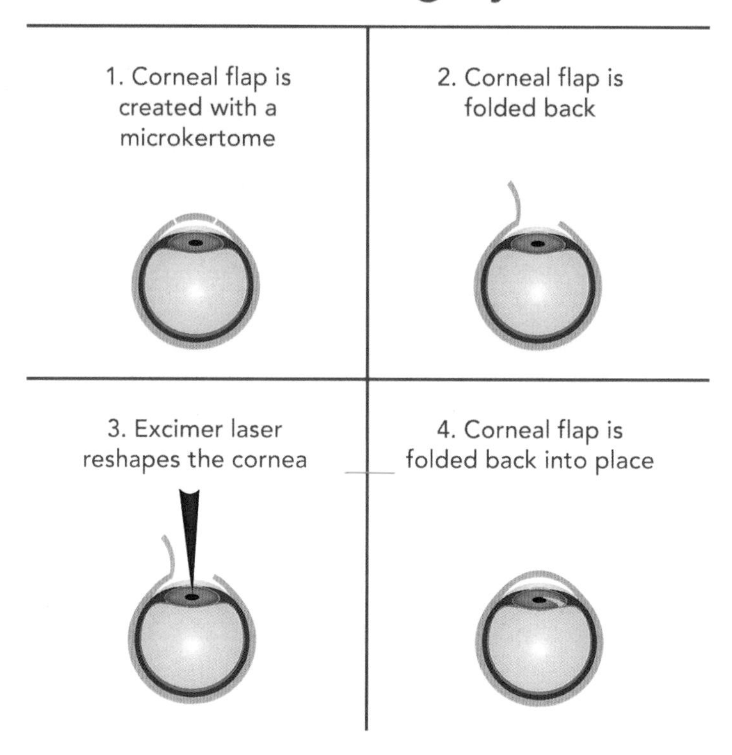

Fig. 4.7 LASIK eye surgery (Adam Chen)

flap complications can cause optical aberrations and distorted vision. Studies indicate that flap complications occur in 0.3–5.7% of LASIK procedures, according to the April 2006 issue of *American Journal of Ophthalmology*. In a study of 3009 consecutive LASIK surgeries performed August 2002–July 2009 using a femtosecond laser for flap creation, flap complications occurred in fewer than one-half of 1% (0.37%) of these procedures, and all complications were successfully managed within the same surgical session. Again, remember that you can reduce your risk of LASIK complications by choosing a reputable, experienced eye surgeon.

3. Irregular astigmatism—This is caused by an unequally curved corneal surface. Irregular astigmatism also can occur from laser correction that is not centered properly on the eye or from irregular healing. Resulting symptoms may include double vision or "ghost images." In these cases, the eye may need re-treatment or enhancement surgery.
4. Epithelial ingrowth—This is when cells from the outer layer of the cornea (epithelium) grow under the flap after LASIK surgery. In most cases, epithelial

ingrowth is self-limiting and causes no problems. But in some cases (reported to be 1–2% of LASIK procedures), symptoms of discomfort and/or blurred vision can occur, and additional surgery is needed to lift the flap and remove the epithelial cells.

5. Diffuse lamellar keratitis (DLK)—With the nickname "The Sands of the Sahara," this is inflammation under the LASIK flap that may have several causes. Some inflammation of the cornea after LASIK surgery is normal. But if it is uncontrolled, as in DLK, it can interfere with healing and cause vision loss. If DLK occurs, it usually responds to therapies such as antibiotics and topical steroids. Also, the flap might need to be lifted and cleaned for removal of inflammatory cells and to prevent tissue damage.
6. Keratectasia or keratoconus—This is a very uncommon bulging of the eye's surface that can occur if too much tissue is removed from the cornea during LASIK or if the cornea prior to LASIK is weak as evidenced from measurements of the cornea. Rarely does keratoconus develop after LASIK with no known risk factors. Enhancement laser surgery is usually not suitable, and gas permeable contact lenses or corneal lenses may be prescribed to hold the cornea in place, or a treatment called corneal collagen cross-linking may be performed to strengthen the cornea.
7. Dry eyes after LASIK—Some people who have LASIK surgery experience a decrease in tear production that can cause eye discomfort and blurred vision. Almost half of all LASIK patients experience some degree of temporary dry eye syndrome, according to the April 2006 issue of *American Journal of Ophthalmology*. Dry eye syndrome after LASIK surgery usually is temporary and can be effectively treated with lubricating eye drops or other measures. Dry eye problems usually disappear when healing of the eye is complete, which can take up to 6 months. People who already have severe dry eye usually are eliminated as LASIK candidates.
8. Significant undercorrection, overcorrection, or regression—Not everyone will achieve 20/20 vision after LASIK eye surgery, and contact lenses or corrective eyeglasses for some or all activities may still be required in rare cases. If the laser removes too much or too little corneal tissue, or the eye's healing response is not typical, the visual outcome will be less than optimal. One possible cause of a less-than-perfect outcome is that the eyes did not respond to laser eye surgery in a predictable manner. Another possible cause is that the eyesight may have been optimal shortly after LASIK but regressed over time due to "overhealing." In most cases, a significant undercorrection or regression can be successfully treated with additional laser vision correction after your surgeon confirms your residual refractive error is stable. Regression of the LASIK effect is common.
9. Eye infection—Infections rarely occur after LASIK. Because the corneal flap acts as a natural bandage, eye infections occur less frequently after LASIK than after flap-free corneal refractive procedures like PRK. Still, it is very important to use medicated eye drops as directed after the LASIK procedure to avoid infection and control inflammation as the eyes heal.

Astronomers should approach LASIK with caution. At night, when the exit pupil is relatively large, the diameter of the exit pupil is equal or exceeds the diameter of the circle of the scarred tissue, causing diffraction/halo artifacts around any bright lights (car lights, street lights). However, once sufficient light is available and the exit pupil shrinks, the artifacts are gone.

LASIK Surgery Incompatibility with Cataract Surgery

LASIK surgery and cataract surgery are two different procedures to improve vision by correcting two different problems: two different roads to the same goal. Lasik seeks to correct focusing disorders like nearsightedness, farsightedness, and astigmatism by reshaping the cornea of the eye. The presence of a small amount of cataract would not diminish the results of LASIK, but if and when the cataracts grow, cataract surgery is still required. This would mean two surgeries on each eye rather than one. Additionally, LASIK is not usually covered by insurance as it is cosmetic, and is considered elective surgery.

One of the unfortunate realities of cataract surgery in an eye that has had previous LASIK surgery is that the process of selecting an artificial lens isn't quite as precise as for those who have not had LASIK. Ophthalmologists have a number of different compensatory methods, but have varying degrees of success with all of them. If the data is available, one tactic used by ophthalmologists is to average the results of four different tests, two of which take into account pre-LASIK measurements, and two based on current measurements post-LASIK.

Cataract surgery replaces the focusing lens of the eye that has become clouded due to aging or other causes. The shape of the cornea is not significantly changed as in the LASIK procedure. Cataract surgery is covered by most insurance providers once the cataract has reached the definition of mature, which can vary.

To add to the level of confusion, some people who desire LASIK are not good candidates for LASIK and can be helped by removing their lens and inserting an intraocular lenses, namely, cataract surgery without the cataracts. This is not customarily covered by insurance providers as it is considered cosmetic in nature.

The Role of Sunglasses and UV Radiation

Unlike the recommendation of the ZZ Top song "Cheap Sunglasses," spend some good money on good quality sunglasses. And wear sunglasses in all seasons, not just the summer. The Sun is emitting a full spectrum of radiation all the time, including the winter.

Ultraviolet (UV) and to a lesser extent blue light are on the high-energy end of the visible spectrum, and as such are contributors to failing eyesight. Ever wonder why, in old movies about the Old West or the desert, blind people are depicted with cataract-laden eyes?

UV rays have higher energy than visible light rays, which makes them capable of producing changes in the skin that create a suntan. UV radiation, in moderation, has the beneficial effect of helping the body manufacture adequate amounts of vitamin D.

Too much exposure to UV causes a painful sunburn. Too much UV radiation and skin cancer becomes an issue. These rays also can cause sunburned eyes—a condition called photokeratitis or snow blindness.

Sunglasses that block 100% of UV are essential to protect the eye from damage that could lead to cataracts and snow blindness.

There is a growing concern over the amount of high-energy blue light (HEV) that enters the eye. The fact that blue light penetrates all the way to the retina (the inner lining of the back of the eye) is important, because laboratory studies have shown that too much exposure to blue light can damage light-sensitive cells in the retina. This causes changes that resemble those of macular degeneration, which can lead to permanent vision loss. Research is being conducted over the concern that the added blue light exposure from computer screens, smartphones, and other digital devices might increase a person's risk of macular degeneration later in life.

Blue light is also good. It's well documented that some blue light exposure is essential for good health. Research has shown that high-energy visible light boosts alertness, helps memory and cognitive function, and elevates mood.

In fact, something called light therapy is used to treat seasonal affective disorder (SAD), a type of depression that's related to changes in seasons, with symptoms usually beginning in the fall and continuing through winter. The light sources for this therapy emit bright white light that contains a significant amount of HEV blue light rays.

Also, blue light is very important in regulating circadian rhythm, the body's natural wakefulness and sleep cycle. Exposure to blue light during daytime hours helps maintain a healthful circadian rhythm. Too much blue light late at night (reading a novel on a tablet computer or e-reader at bedtime, for example) can disrupt this cycle, potentially causing sleepless nights and daytime fatigue.

Anyone who spends a lot of time outdoors is at risk for eye problems from UV and HEV radiation. Risks of eye damage from UV and HEV exposure change from day to day and depend on a number of factors, including:

1. Geographic location—UV levels are greater in tropical areas near the earth's equator. The greater the distance from the equator, the smaller the risk.
2. Altitude—UV levels are greater at higher altitudes.
3. Time of day—UV and HEV levels are greater when the sun is high in the sky, typically from 10 a.m. to 2 p.m.
4. Setting—UV and HEV levels are greater in wide open spaces, especially when highly reflective surfaces are present, like snow and sand. In fact, UV exposure can nearly double when UV rays are reflected from the snow. UV exposure is less likely in urban settings, where tall buildings shade the streets.
5. Medications—Certain medications, such as tetracycline, sulfa drugs, birth control pills, diuretics, and tranquilizers, can increase your body's sensitivity to UV and HEV radiation.

Surprisingly, cloud cover doesn't affect UV levels significantly. The risk of UV exposure can be quite high even on hazy or overcast days. This is because UV is invisible radiation, not visible light, and can penetrate clouds.

There are various eyeglass formulations and sunglasses that filter UV radiation and provide degrees of filtration of blue light frequencies. Again, consult your ophthalmologist and optometrist on the best choices.

And don't wear sunglasses at night … it's not cool. It's silly!

Except when an eyeglass wearer has photochromic lenses in their glasses. Photochromic lenses are eyeglass lenses that are clear indoors or at night, and darken automatically when exposed to sunlight. Amateur astronomers who wear eyeglasses should consider these for everyday wear. There is a convenience factor and safety element to always having UV protection for eyeglass wearers. Some people like these, other don't. The big complaint is that they're either too dark indoors or not dark enough outdoors.

The molecules responsible for causing photochromic lenses to darken are activated by the sun's UV radiation. Because UV rays penetrate clouds, photochromic lenses will darken on overcast days as well as sunny days.

Photochromic lenses typically will not darken inside a vehicle because the windshield glass blocks most UV rays. Recent advancements in technology allow some photochromic lenses to activate with both UV and visible light, providing some darkening behind the windshield.

Chapter 5

It is clear to everyone
that astronomy at all events compels the soul to look upwards,
and draws it from the things of this world to the other.

—Plato

There is an old amateur astronomy adage that states: *The telescope that gets used the most is the telescope that sees the most.* Meaning a smaller telescope that's easy to move and set up gets used more often and therefore sees more of the sky than a large telescope that gets used only occasionally. This adage is in direct conflict with another old astronomy concept: *Aperture Rules.*

At different stages of an amateur astronomer's life, a different set of telescope equipment is optimal. At the age of 20–30+ years old, anything goes. Large aperture telescopes with heavy cumbersome mounts are not a problem. Any eyepiece that is the best in a particular performance area is purchased, collected, and used.

But as one ages, the formula for astronomical success shifts. The best way to examine this is to study some scenarios.

J.L. Chen, *Astronomy for Older Eyes*, The Patrick Moore Practical Astronomy Series, DOI 10.1007/978-3-319-52413-9_5

Single Large Telescope at Age 70+

This first case study is that of a 75 year old man, let's call him John, who owns a large Celestron C-14 Schmidt–Cassegrain Telescope (SCT) on a huge Celestron CGE German equatorial mount. John's owned this setup for years, and when he was younger this was the telescope setup that he was proud to own and use. The telescope and its mounting weigh over 199 lbs, not including the associated accessories of eyepieces, diagonals, batteries, etc. When John was younger, he was able to assemble, set up, and align this mammoth telescope from its storage in the garage in less than 20 min. But now at age 75, this is not necessarily the case. The telescope disassembles into three awkward pieces, with the telescope mount itself weighing a whopping 75 lbs. And even though the Optical Tube Assembly (OTA) is the lightest component at 45 lbs, it is awkwardly placed at the top of the tripod/pier and equatorial mount assembly at a height of 50+ in. off the ground. Added to this are 22 lbs of counterweight. At 75 years old, assembling this behemoth becomes a daunting task. As a result, the telescope gets used less and less over time (Fig. 5.1).

Fig. 5.1 Celestron founder Tom Johnson next to a Celestron C-14 (Celestron)

John's eyepiece collection also has not stood the test of time. His collection of high-quality Plossl and Abbe orthoscopics worked well when he was younger, but now he has to wear glasses. With glasses on, he can't see the full field of view because of the lack of eye relief in the older eyepiece designs.

John is slowly losing what was once his lifetime passion.

Multiple Telescopes at Age 64

This next case study is that of a 64-year-old man, named Jim (guess who this is based on!?!), who owns a moderately large Celestron 11 SCT on a GoTo fork mount, a 102 mm apochromatic refractor on a GoTo mount, and an 80 mm refractor on an alt-az mount. Jim is barely able to move the 11-in. SCT with its tripod attached (weight 92 lbs assembled) eight steps from inside his family room through the sliding glass door onto his deck. The 11-in. was acquired while Jim was still in the workforce and years away from retirement. The high-quality 80 mm has long been a member of Jim's stable of telescopes to fill the need of a grab-and-go scope. The 102 mm refractor was acquired shortly after Jim's early retirement in his mid-50s as a quality optical telescope that was easily transported to star parties. The assembled 102 mm refractor on its mount is a tidy 38 lbs and easily transported to the deck, or packed into the car for star parties. Jim's 80 mm refractor is the "grab-and-go" star of the collection, which Jim can grab a quick look at the Moon or Jupiter in between scattered clouds, or for an impromptu camping trip. The least used telescope of Jim's collection is the 11-in. SCT, for those one or two perfectly clear nights per month where good weather is forecasted and extended observing is planned. However, the 102 mm and 80 mm receive 90% of Jim's observing time, due to their portability and ease of setup (Fig. 5.2).

Jim uses modern design eyepieces that produce a wide-field, high-contrast sharp image with 20 mm of eye relief, a must for he also wears glasses.

Telescope in an Observatory at Age 68

This case examines the case of Charlene, who has built a full functioning observatory, with a computer GoTo 16-in. telescope, 4-in. photo guidescope, CCD imaging equipment, and high-tech computer hardware and software to fully control all functions of the observatory, including programmed imaging sessions where Charlene does not need to be present. All Charlene's equipment is permanently installed in the observatory, and all instruments can be completely controlled remotely. All this is located in a deep sky site. However, there is no portability (Fig. 5.3).

When Charlene decides to do some visual observing, she has a different collection of eyepieces depending on what she is observing. High contrast eyepieces are for planetary observing. Ultra-wide field eyepieces are used for extended deep sky objects. With the CCD imagers at hand, Charlene also does a lot of observing

Fig. 5.2 80 mm and 102 mm refractors and 11-in. SCT (James Chen)

Fig. 5.3 Meade LX-200 16-in. SCT installed in a dome (Hands-on-Optics archive)

from the comfort of his home and receiving and viewing the telescope images from her laptop.

But Charlene's ophthalmologist has told her that her eyeglass prescription is changing, and she is in early stages of cataracts.

Telescopes, Aging, and Physical Strength

In the cases of John and Jim, the physical strength of their youth has given way to the ravages of time. Large telescopes that were manageable in their 30s and 40s are no longer manageable in their 60s and 70s. Additionally, the ease of use has become a major factor.

According to muscular strength studies conducted by Karsten Keller and Martin Englehardt, and reported in the Muscle, Ligaments, and Tendons Journal in 2013, the aging process is connected with widespread and typical changes in the human body. With increasing age, body composition is changes with a loss of muscle mass and bone mass and a reduction of physical capacity over years.

Maximum physical capacity occurs between ages 20 and 30 years. For those ages between 30 and 50 years, the published changes in muscle mass, power, and strength are small. Pronounced changes due to the aging process take place after age 50. Profound changes in muscle strength occur with the lean body mass of the legs losing of 1–2% per year and a leg strength loss of 1.5–5% per year are reported for individuals older than 50 years.

The study concluded that the aging process leads to a distinct muscle mass and strength loss. The decline of the muscle strength of people, who were younger than 40 years, in comparison to those, who were older than 40 years, ranged between 16.6% and 40.9%.

Strength peaks around 25 years of age, plateaus through 35 or 40 years of age, and then shows an accelerating decline, with an average 25% loss of peak force by the age of 65 years. The mechanisms underlying the aging process are not well understood. Possible hypotheses include a "wear and tear" which exceeds the reparative capacity of the tissues, a development of immunity to the individual's own protein constituents, and errors in cell division, associated with exposure to external radiation or endogenous mitogens such as peroxidases. Some biologists have even argued that aging has been "programmed" by evolution to avoid the hazard of overpopulation.

Loss of strength progressively impedes everyday living. It becomes difficult to carry a 10 pound bag of groceries, to open a bottle of medicine, and even to exit a car seat. The male/female strength ratio is unchanged, and women are limited by a loss of strength at an earlier age than men.

Muscle strength can be greatly improved with as little as 8 weeks of resistive training. Muscle mass building proceeds more slowly than in a 20–30-year-old adult, but cross-sectional comparisons between active and inactive individuals suggest that much of the wasting of lean tissue can be avoided by regular resistive exercise. The resistive training enables a slowing of the degenerative process of aging, but does not stop it.

In the case of both John and Jim, regular exercise that includes strength training will help them handle their oversized telescopes. But the path that Jim has chosen, and that John should seriously consider, includes the use of smaller and more user-friendly telescopes. In terms of light gathering and resolution, these smaller telescopes cannot match the overall performance of John and Jim's big eye telescopes, but the more frequent use of these user-friendly telescopes cannot be denied. In the case of John, there are a number of smaller SCTs available in the 5–8 in. aperture range that are light, easily manageable, high-tech with GoTo capability and weigh less than 30 lbs. The mass of big eye telescopes is the instant cure for aperture fever when the birthday number 50 rolls around.

An alternative option available to both John and Jim is to equip their large telescopes with wheels or JMI wheely bars. No lifting of the equipment is involved, just a careful push of the telescope out a door or garage as it rolls on wheels onto a patio or driveway.

The ultimate option available to John and Jim is the solution that Charlene has chosen, a dedicated observatory to house their big telescopes. This is not necessarily the most economical option, nor does it address the problem of transporting the telescope to remote viewing sites or star parties. As Charlene grows older, the maintenance of the observatory, including painting, waterproofing, and cleaning, will become tasks that will not be handled as the observatory falls into disrepair.

In general terms, for good overall health reasons, John, Jim, and Charlene still need physical exercise and physical activity, no matter whether an astronomer or not.

Telescope Eyepieces and Aging Eyes

In all three scenarios, John, Jim, and Charlene all share a common dysfunction, slowly failing eyesight and the need for wearing corrective lenses in the form of glasses, bifocals, or contact lenses. Presbyopia has affected all three (Fig. 5.4).

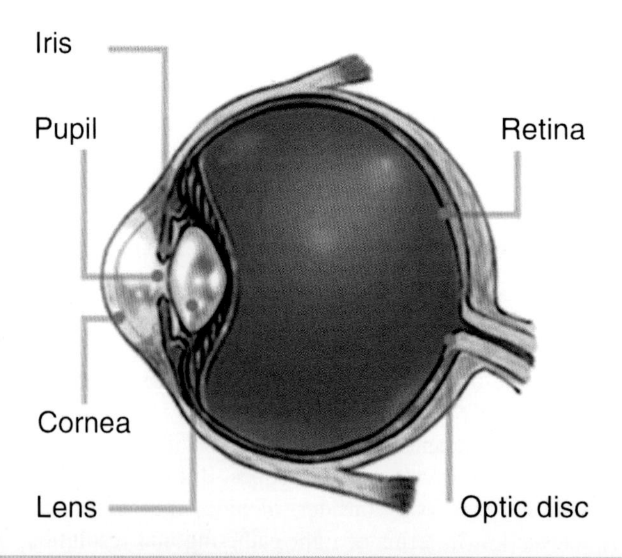

Fig. 5.4 Anatomy of the Eye (NIH)

Normal Age-Related Vision Changes

A major age-related eye change that affects an amateur astronomer is that the pupil becomes smaller and less responsive to variations in light. A teenager's pupil is capable to opening to at least 7.0 mm, but with age that pupil opening falls into the 5.0–5.5 mm.

Because the pupil controls the amount of light that reaches the retina, age-related changes to the pupil may affect vision in many ways. First, as the pupil decreases in diameter, seeing well in dim light becomes harder. In addition, the less able the pupil can adjust to varying light conditions, the less tolerable glare becomes and the more difficult it is to adapt from darkness to bright light or vice versa. This means as you get older, you may need more time to adjust to changing levels of illumination, such as going from bright sunshine into a dimly lit room or restaurant.

This is where the exit pupil for a particular eyepiece in a specific telescope is important. The exit pupil is the diameter of the beam of light leaving the eyepiece and traveling to the observer's eye, where it enters the observer's own eye pupil.

There are two simple formulas that can be used to calculate the exit pupil for a particular eyepiece on a specific telescope. The first formula is:

$$\text{Exit Pupil} = D/M$$

where

D = the diameter of the telescope's objective lens or primary mirror in millimeters

M = magnification = focal length of telescope/focal length of eyepieceor

$$\text{Exit Pupil} = F/f$$

where

F = the focal length of the eyepiece in millimeters

f = the telescope's focal ratio (the *f*-number)

So technically, what does this mean? In a nutshell, it means as one ages, a person's light sensitivity for low power and wide-field viewing slowly diminishes with age. When young, an individual's pupil will open to 7 mm, and a telescope and eyepiece combination that produces an exit pupil of 7 mm will produce a bright wide image. However, with that same telescope/eyepiece combination, a 70-year-old with a pupil that opens only to 5.5 mm will see a dimmer and somewhat less spectacular view. The 70-year-old pupil is just not physically capable of opening to the degree as that of the 20-year-old.

Another age-related eye change is the lens of the eye begins to lose elasticity. In the same way that losing flexibility in tendons and muscles makes it more difficult for the body to move, losing lens elasticity also makes it harder for the lens to bend in order to focus on closely held objects. This loss of focusing power, or lens accommodation, is known as presbyopia. Presbyopia is the reason that bifocals and reading glasses were invented.

The challenge of presbyopia on an amateur astronomer is two-fold. The first challenge is focusing the telescope properly. This is where the owner of a long focal length telescope of *f*/15 has an advantage over the *f*/5 telescope owner. As in photography, the longer the *f*/ratio, the greater the depth of field. The owner of the *f*/15 telescope is able to bring the image in the eyepiece into focus much easier because of the greater leeway the long focal length allows. The sharper and steeper *f*/5 light cone that enters the eyepiece allows for much less error in focusing. Add to this discussion that an aging eyeball compounds the focusing challenge.

Those familiar with the telescope market have seen a proliferation of dual speed focusers that allow course focusing with one knob and a 1:10 reduction in focus speed with the other focusing knob. Originally intended for astrophotographers, these dual speed focusers are a must-have for older amateurs (Fig. 5.5).

An additional age-related change in the eye is the gradual yellowing of the lens of the eye with age. The yellowing of the eye lens affects color perception. For example, the yellowing lens tends to absorb and scatter blue light, making it difficult to see differences in shades of blue, green, and violet. Colors may seem duller, and contrasts between colors will be less noticeable. This particularly affects those amateur astronomers that specialize in planetary observations. The viewing of the subtle differences of the cloud bands of Jupiter becomes a challenge. Mars, a notoriously low contrast planet to observe at opposition, becomes even harder to view satisfactorily. Unfortunately, the yellowing of the lens is often the precursor to cataracts.

The aging astronomer does have filter technology to help distinguish details through the eyepiece. There are a number of filters available to telescope owners, such as nebula filters, light pollution filters, and color filters. These filters are very

Fig. 5.5 The dual speed focuser on the author's 102 mm refractor (James Chen)

Fig. 5.6 A color filter set (Hands-on-Optics)

useful in many applications where the goal is to reveal very dim low contrast objects and features (Fig. 5.6).

Eyepiece color filters work by blocking out certain colors in the visible spectrum of light. A red filter, for example, will block out all but the red wavelength of light. If you look at an object that is primarily red while using a red filter, the object will appear very bright. Areas which are not red will appear more clearly because they contrast with the wavelength of light which is being passed by the filter. Filters are sorted by the Kodak Wratten numbering system. Each filter is listed by its color and Wratten number. The Wratten numbers will help to ensure similar results between different filters. The image should appear the same when viewed through any #82A Light Blue Filter, for example. Details of color filter usage are located in Appendix 2.

The choice of telescope eyepieces when the individual is an eyeglass wearer is important. For critical observing, it is best to remove the eyeglasses, since the optical quality of eyeglass lenses is not to the same level of smoothness and accuracy as the optics of the telescope or eyepieces (Fig. 5.7).

Eye relief is the distance from the eye lens to the observer's eye where the entire field of view can be seen at once.

For the matter of convenience and when sharing the telescope views with others without continual refocusing, the ability to see the full field of view while wearing glasses is important to many backyard astronomers. Eye relief is a function of the particular eyepiece design. Generally speaking, many modern eyepieces designed within the past 20 years will have good eye relief. Older classic designs, such as Abbe and Plossl orthoscopics, tend to offer eye relief of only ¼–½ times the eyepiece's focal length. Classic eyepieces become very uncomfortable to view, especially in the shorter focal lengths. The design goal of many modern eyepiece designs is to extend the eye relief to 18–20 mm at all focal lengths, thus enabling

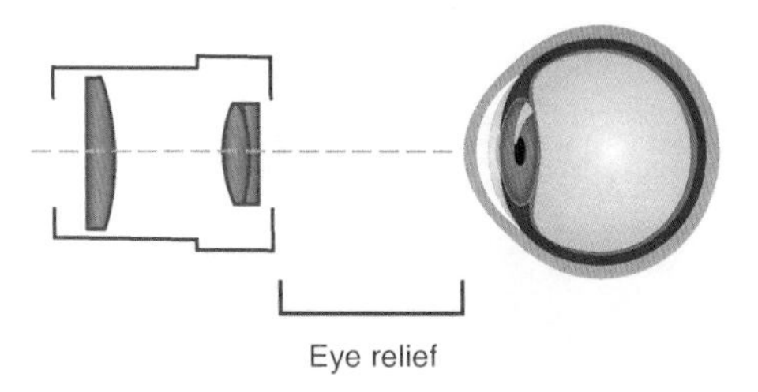

Fig. 5.7 Eyepiece and eye relief (Adam Chen)

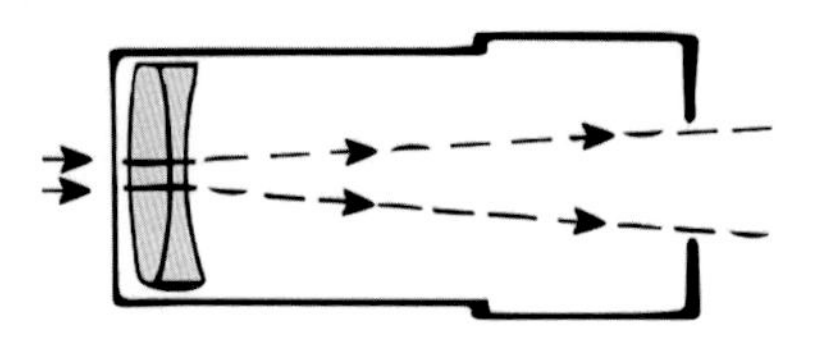

Fig. 5.8 The Barlow Lens design (Adam Chen)

eyeglass wearers the luxury of viewing with their eyeglasses on. The increased eye relief is also enjoyed by the non-glass wearer too. There are some eyepiece designs that claim long eye relief, but are only 13 to 15 mm of eye relief. Eyeglass wearers may find this a little tight. The advice here is try before you buy.

An alternative to buying the high eye relief eyepieces that can be expensive, it is possible to obtain excellent eye relief by using a longer focal length (low power) Plossl or Abbe design with a 2× or 3× Barlow lens.

For the uninitiated, a Barlow lens is a useful addition to every eyepiece case. A Barlow lens is a negative lens system placed along the light path between the objective and the eyepiece that increases the effective focal length of the telescope, therefore increasing the magnification. Typically, Barlows double (2×) or triple (3×) the magnification. Newer focal extenders using three or four lens elements are available to quadruple (4×) or quintuple (5×) the focal length. These accessories are useful in three ways. A single Barlow lens effectively doubles the number of magnifications available in an eyepiece collection. The use of a focal extender also allows longer focal length eyepieces with their higher eye relief to be used at higher magnifications for eyeglass wearers. The use of a Barlow lens can improve the off-axis edge sharpness of some eyepiece designs. The Barlow lens and related focal extenders are also useful for astrophotography (Fig. 5.8).

High Tech Versus Low Tech

Telescope optics technology has advanced over the years without greatly impacting the basic operational use of the telescope. Despite the changes in optical materials and configurations, the basic concept of a telescope objective gathering light and

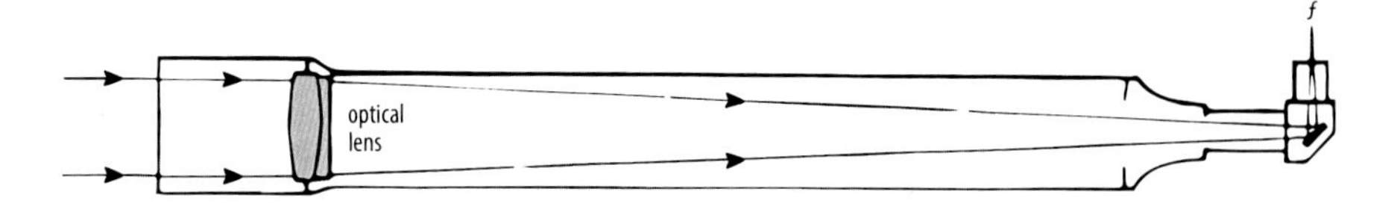

Fig. 5.9 The refractor (Adam Chen)

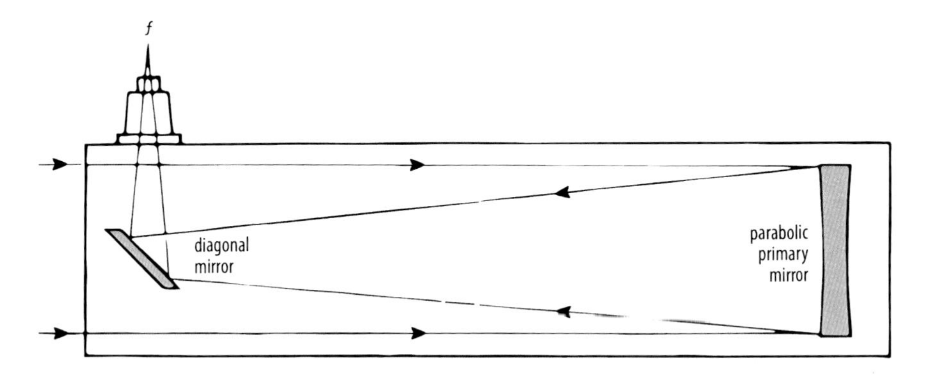

Fig. 5.10 The Newtonian reflector (Adam Chen)

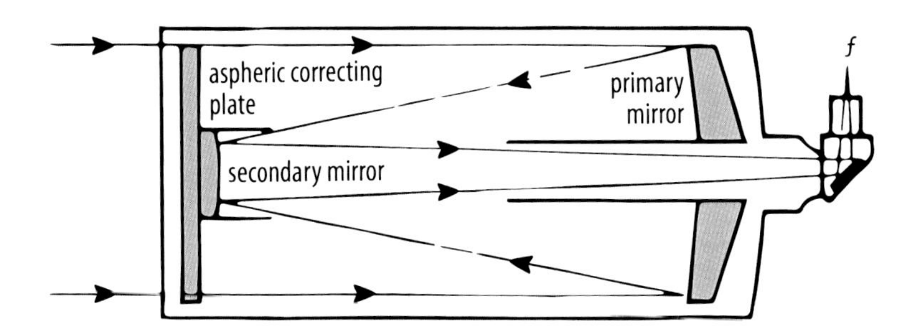

Fig. 5.11 The Catadioptric—Schmidt–Cassegrain telescope (Adam Chen)

the eyepiece positioned and focused to produce an image has not changed (Figs. 5.9, 5.10, 5.11, and 5.12).

As discussed in Chap. 4, the refractor telescope, by virtue of having an unobstructed optical light path, has a contrast advantage over reflector and catadioptric telescopes. For those who already own an SCT or Newtonian telescope, the difference in performance does not necessitate selling the old telescope and buying a refractor strictly on the basis of optical performance. Other issues, to be discussed further, such as portability and ease of use will play a larger factor as you get older.

For beginners, beware of the "beginner" 4½ in. reflector telescope. Many sources recommend as the ideal beginner telescope the 4½ in. Newtonian telescope,

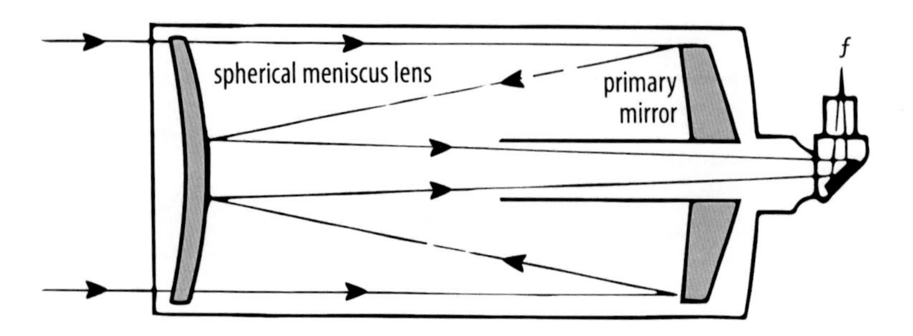

Fig. 5.12 The Catadioptric—Maksutov–Cassegrain telescope (Adam Chen)

many offered with low acquisition prices. These recommendations are generally based on a low price per inch of aperture, with a heavy emphasis on the adage "Aperture Rules." This is not a good idea. What's often missing from these recommendations is the fact that Newtonian telescopes require regular maintenance, i.e., optical realignment. The anecdotal evidence from telescope retailers is that up to 90% of all telescopes in their backrooms awaiting repair are 4½ in. Newtonians and Dobsonians (same telescope, different mountings) that simply need to be aligned because their owners lack the skill or knowledge to properly set up a reflecting telescope.

For those buying their first telescope, with all things being equal, the 4-in. refractor is highly recommended for optical quality and ease of use. The 4-in. refractor offers similar light gathering (remember, no central obstruction that blocks incoming light and compromises contrast), superior contrast, and needs no adjustments.

A solid telescope mount completes the total system needed for viewing the Moon, and beyond. There are two basic flavors of mounts: the altitude-azimuth mount, mostly referred to as the Alt-Azimuth or AltAz mount; and the equatorial mount. Each type can come either as manual, driven by hand controls or motors, or computer-driven GoTo models. A complete discussion of these types of mounts can be found in Appendix 1.

Of particular interest to older astronomers is the new technology of GoTo technology. Advanced electronic technology has enabled the proliferation of GoTo computerized telescopes. A GoTo telescope mount is quite simply a telescope system that is able to find celestial objects in the night sky, and then track them. The GoTo mount can be set up in an alt-azimuth or equatorial fashion, and after the proper alignment procedure, the finderscope is no longer needed for the rest of the evening. Some of the newer GoTo telescopes have electronics and CCD cameras that will perform the alignment procedure automatically.

These telescope mounts are wonderful pieces of modern engineering design. The GoTo technology allows for more efficient use of observing time by quickly finding objects in the night sky. Built into the hand controller is a microprocessor, firmware, and built-in memory catalog of the positions of thousands of stars, galaxies, nebulae, open star clusters, globular clusters, planetary nebulae, our solar system planets, and the Moon. Complex algorithms developed and refined over the years with improvements in encoders and motor design have made the GoTo telescope an accepted and desirable telescope feature. The more sophisticated designs feature a planetarium-type display with convenient easy-to-use pull-down menus for ease of operation. The use of these planetarium-type GoTo mounts is featured in this author's works *The Vixen Star Book User Guide* and *The NexStar Evolution and SkyPortal User Guide.*

There is an ongoing debate within the amateur astronomy community on the merits of computer-guided and computer-controlled telescopes. The hardcore conservative backyard astronomers argue that a beginner or novice individual is better served learning the skies without electronic aids, as generations of stargazers have done. There is merit to this argument. However, in these days of increasing light pollution in urban and suburban neighborhoods, seeing landmark stars used for "starhopping" to locate deep sky objects is becoming increasingly difficult and frustrating to a backyard astronomer, particularly to the beginner or novice. Using bright first magnitude stars for alignment, a computerized GoTo system eliminates frustration and introduces fun into the hobby. The search time for a celestial object is reduced from tens of minutes to mere seconds! With the electronics aiding the observer in finding the deep sky objects, a suburban observer can then take advantage of modern filter technology in overcoming the light pollution in their area. Cheers to the miracle of nebula filters, light pollution filters, and color filters!

Of course, in the worst of urban environments, even using a GoTo telescope and mount can be challenging, especially if bright stars are impossible to see for alignment purposes or otherwise. For instance, in the middle of brightly lit Las Vegas, the only bright stars visible are Wayne Newton, Celine Dion, and a variety of Elvis impersonators!

The problem with GoTo telescopes for the older amateur astronomer is their ability to learn and use the computerized hand controller of a GoTo mount. The simple decision point comes to this: If the microwave clock is flashing 12:00 and the use of an iPhone causes confusion, the older individual should stay away from the GoTo telescope and just use a manual or motor-driven telescope mount. If the older amateur astronomer is comfortable using a laptop computer, owns and uses smartphones and tablets, can program their television remote to control their TV, cable, and DVR, they should easily be able to understand the complexities of a computerized hand control and a GoTo mount.

Astrophotography and its modern cousin astro-imaging is a different subject altogether. Astrophotography, either with film cameras or with CCD cameras or DSLRs, offers a major challenge to all comers, young and old alike. The knowledge and skills far exceed that of normal point-and-shoot photography. Post-processing of digital images requires a considerable amount of computer knowledge, and this often exceeds the comfort zone of many older backyard astronomers. The best advice here is for someone new to astronomy to hold off astrophotography for 2 or 3 years to get accustomed to astronomy as a whole before even attempting their first astro-photo. Fall in love with astronomy first, then try astrophotography. If you find astrophotography too much of a challenge, the love of observing will still be there. The backrooms of many telescope stores are filled with equipment from those who failed at astrophotography and did not have the love of observing as a fallback position.

The Role of Binoculars

An often overlooked piece of equipment in the amateur astronomer's arsenal is the binocular. Many newcomers to astronomy are frequently told to begin their study of the night sky with a pair of binoculars. Backyard astronomers without GoTo telescope mounts often use a pair of binoculars as part of their search for night sky objects before zooming in with their telescopes (Fig. 5.13).

For the newcomers to astronomy, the appeal of the binocular needs some explanation. Assuming the newcomer was drawn to astronomy because of attending an outreach star party or a viewing through a friend's telescope, the excitement of astronomy was seeing Saturn's rings or the Great Orion Nebula at medium to high magnifications. Seeing those kinds of thrills will be missing with 7 × 50 or 10 × 50 binoculars. The newcomers to astronomy, armed with a newly acquired telescope, will find themselves in the uncomfortable position of being totally unfamiliar with the night sky and will have difficulty in finding astronomical targets. Enter naked eye and binocular viewing. Armed with star atlases, or for the more technically oriented iPhone or iPad app, and binoculars, the newcomer becomes familiar with

Fig. 5.13 The author's binocular collection (James Chen)

the constellations, the Milky Way, and star fields. Binoculars used in tandem with a telescope ease the learning curve of the neophyte backyard astronomer.

For older observers, the binocular is a convenient, grab-and-go optical tool that allows for wide-angle viewing of star fields, large open clusters, locating tight globular clusters, and identifying large craters and dark mares regions on the Moon. So long as the older amateur astronomer is not affected with Parkinson's Disease or similar medical conditions that cause the hands to tremble, an amazing amount of night sky observing can be done with binoculars.

Experienced amateurs, having already completed their Messier, brighter NGC objects, and Caldwell objects lists with their telescopes, find a new challenge in binocular observing and searching the familiar objects from a lower magnification and wider field perspective. Seeing the object and its relationship to the sky around it will put that object in its proper context in the sky.

Binocular Basics

For astronomy, the old rule of thumb is aperture is king and light grasp is the most important factor. Choose binoculars with at least a 50 mm aperture. A 7 × 50 model is most often the recommended binocular because it is easily hand-holdable and provides nice, wide-field views of starry swaths. However, age plays a big factor in choosing a pair of binoculars. Since after age 50, the pupil dilates or opens up to about 5.5 mm or smaller, the 7.1 mm exit pupil of the 7 × 50 binocular results in wasted light gathering by the binocular. Another factor affecting the exit pupil is the observing environment. While observing from the urban or bright suburban environment, light pollution will affect the exit pupil and will not allow it to fully dilate as under a country sky. For instance, observing with a 7 × 50 pair of binoculars under a city sky is like observing with a pair of 7 × 35 binoculars. Why? Because the pupil is not fully dilated to take advantage of the extra exit pupil or circle of light being projected by the 7 × 50 binoculars. What is the exit pupil of a 7 × 50 binocular? Answer. 7.1 mm. Divide 50 mm, the aperture of the binocular in this case, by the power of the binocular, 7×. The 7 × 35 binocular exit pupil is 5 mm. To get the most out of a 50 mm aperture, an 8 × 50 or better yet a 10 × 50 binocular, is recommended (Figs. 5.14 and 5.15).

Also, when buying a pair of binoculars, check out the eye relief or how close you have to hold your eye to the binocular, to get the full field of view. Some eyeglass wearers require a long eye relief. Usually eye relief of 15 mm or more is sufficient for eye glass wearers and anyone else.

The higher-power 10 × 50 is also popular, and in fact is preferable to the 7 × 50 where skyglow light pollution is a problem. A tripod is recommended for a steady view, and if viewing through the larger 70–125 mm astronomy binoculars, heavy duty tripods are mandatory. These "Giant" astronomy binoculars of 70 mm, 80 mm, 100 mm, or 125 mm aperture will reveal fainter deep sky objects and more subtle detail. Some even allow interchangeable eyepieces, allowing the user to

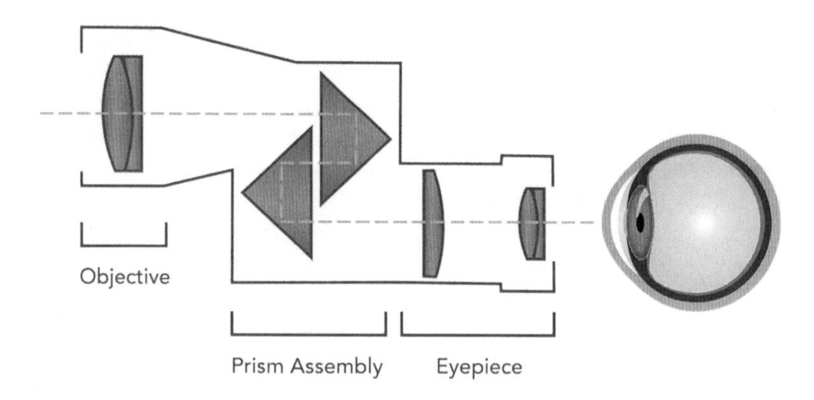

Fig. 5.14 Porro-prism binoculars (Adam Chen)

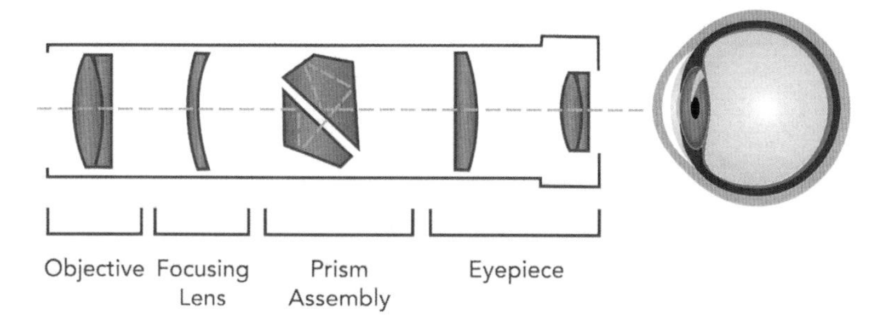

Fig. 5.15 Roof-prism binoculars (Adam Chen)

switch from low wide-field views to a medium power closer-in view. The giant binoculars are not recommended here for older observers. They are big, bulky and as much trouble setting up on tripods as telescopes! Don't think of these binoculars as instruments for high-power viewing. That's what telescopes are for! The giant binoculars are not cheap, and often exceed the price of a similar aperture telescope by thousands of dollars. Remember, think of giant binoculars as twin telescopes.

Another option is obtaining a bean bag chair. By nestling into the bean bag with binoculars in hand, the elbows can be rested on the bean bag sides as they puff up from your weight and provide support to the forearms and the binoculars! An easy and comfortable way to observe the night sky with binoculars. This technique has been used by the author successfully with binoculars up to 15 × 70s.

The most popular models for bird-watching are 8 × 40 and 8 × 42. They're small and nimble, offer steady handheld views, and have sufficient light grasp to provide bright, well-resolved images. These can be used for casual night sky watching.

Binocular Prisms: Porro Versus Roof

The prisms inside binoculars turn what would otherwise be an upside-down image right-side up. There are two main types of prism systems used in modern binoculars: porro prisms and roof prisms.

Porro-Prism Binoculars—Each barrel of a porro-prism binocular contains two right-angle prisms. They are offset from each other, which requires that the objective lenses be spaced farther apart than the eyepieces. Thus, porro-prism binoculars are bulkier than their roof-prism counterparts. Optically, however, porro-prism binoculars usually perform better, because the prism design requires less strict tolerances. That makes them easier to manufacture, so they cost less. Also, porro-prism binoculars yield a more stereoscopic, or three-dimensional, image.

Roof-Prism Binoculars—The prisms in a roof-prism binocular overlap closely, allowing the objective lenses to line up directly with the eyepieces. This results in a more streamlined, compact, and often more lightweight binocular than equivalent porro-prism models. But roof-prism binoculars are also more difficult to manufacture, so they cost more. Roof prisms lose slightly more light to reflection than porro prisms—a disadvantage for astronomical use but not a concern for daytime terrestrial viewing. Well-made roof-prism binoculars can provide optical performance nearly equal to, but not better than, porro-prism binoculars.

Prism Glass—Most optical prisms are made from either BK-7 (borosilicate) glass or BAK-4 (barium crown) glass. BAK-4 is the higher-quality glass and yields brighter images and better edge sharpness. It is also more expensive, but worth the money.

Field of View

The field of view is the size of the "window" as seen through a binocular. The field has no effect on the size of the subject being viewed (that's a function of the magnification). Surprisingly, the field of view is not determined by the binocular's main (objective) lenses, but rather by the eyepiece and prism designs.

The width of the field is expressed either as an angular measure in degrees, or as a linear width measurement, in feet, of the viewing area at a distance of 1000 yards. Conversion from angular field to linear field is accomplished by multiplying the angle in degrees by 52.5.

"Wide-angle" is an inexact term that simply indicates that a binocular field is wider than average. Generally, a binocular is considered wide-angle if its true field of view is 6° or greater.

A wide-angle binocular is ideal for finding subjects quickly, and it can deliver spectacular panoramic views. A wide field is especially desirable for watching action sports or scanning to pick up motion at close range, such as in a wooded area. A narrower field of view is sufficient for longer-distance observation.

Deciding on Binoculars

Deciding which pair of binoculars are best depends on particular goals for using the binoculars. As with telescopes, many amateur astronomers have two or more sizes of binoculars to handle a variety of viewing conditions and targets.

These are important factors to consider:

1. Are there multipurpose roles for the binoculars?—By far the most important factor to consider when choosing a binocular. For daytime outdoor use while hiking, a smaller lightweight, compact and portable binocular is preferable. The hiker's binocular, often a 6 × 25 or 8 × 30 model, will likely be aperture challenged when used for astronomy purposes. Birding enthusiasts and nature lovers find binocular models will find versatility with 8 × 42 or 10 × 42 models, as they provide brighter views with more detail visible in twilight hours and reasonably good light gathering for nighttime views than the smaller hiker-sized binoculars. For binoculars strictly for nighttime use in viewing the nighttime skies, a larger astronomy binocular with at least 50 mm diameter lenses is required for the best views. Many astronomers favor 7 × 50, 10 × 50, 8 × 56, 10 × 60, and 12 × 60 binoculars.
2. Binocular Design: Porro versus Roof Prisms—In general, porro prisms provide brighter views, but are often bulkier than roof prism binoculars. Roof prism binoculars are significantly more compact and portable than porro prism models, but they do not provide the same level of brightness, and are often more costly. Roof prism binoculars are easier to handle. Try before buying to find the right choice.
3. Field of View—A pair of binoculars with a wide field of view allows for scanning vast areas with magnified vision. If the binoculars are anticipated to be used in secondary roles, such as sporting events or scenic landscapes, a binocular with relatively low magnification and a wider field of view is recommended. However, if birding and wildlife observing is combined with astronomy, a higher-magnification binocular with a more narrow field of view may be preferable. Field of view is expressed as feet at a thousand yards for most binoculars. To convert this to astronomical field of view or degrees, use the following formula:

$$FOV\left(\text{in degrees}\right) = FOV\left(\text{in linear feet}\right)/52.5$$

Spend a few dollars extra for a well-designed binocular with Bak-4 prisms, fully multicoated lens, air-spaced objectives, and if available, nitrogen-filled tubes sealed with o-rings to water proof the binocular. Bak-4 prisms, fully multicoated optics, and air-spaced objectives allow for better light transmission, and therefore, a better view. Air-spaced objectives allow for better resolution. The nitrogen-filled tubes sealed with o-rings keeps water, whether of the liquid variety or the vapor variety, from entering the binoculars system and causing mold or mildew to grow.

Of major importance are binoculars that are in collimation. Collimation means all the optics for the right side of the binocular are in alignment with the left side of the binocular. Looking through a well-collimated binocular will produce a view that is perfectly comfortable with no eye strain or double vision. Binoculars are unusable when the lenses' optical axes are not lined up with the mechanical axes of the tubes correctly. For a quick collimation test, do the following:

1. Spot an object 50–100 yards away and focus for both eyes.
2. Look at the target and take the binocular down and let your eyes relax for half a minute.
3. Look at the target again, but cover the right objective with your hand.
4. Pull your hand away from the right objective.
5. If the target is out of focus, but quickly goes back into focus, your eyes are adjusting to the inherent error in collimation, and the binocular should be rejected.

Figures 5.16 and 5.17 are the extremes for handheld binoculars. The 12 × 60 binoculars are at the upper range of what can be easily handheld, but providing the

Fig. 5.16 Celestron 12 × 60 binoculars (Celestron)

Fig. 5.17 Vixen 2.1 × 42 binoculars (Vixen)

light gathering of 60 mm objectives and a 5 mm exit pupil useful for older observers. The 2.1 × 42 binoculars are at the opposite extreme, being small, lightweight, pocket-sized, and extremely wide field in providing a 25° field of view. Large star fields come to life with the 2.1 × 42, giving the feeling of having 42 mm eyeballs. Deep sky objects such as globular clusters, open clusters, and the Andromeda Galaxy are easily identifiable and set in their stellar surroundings. The effect is exhilarating.

For purposes of an older observer, this duo of binoculars is highly recommended.

Solar Observing

One oft-overlooked form of amateur astronomy is observing our nearest star, the Sun. Solar observing requires special equipment to view the Sun in safety.

IMPORTANT SAFETY INFORMATION: Observing the Sun should only be attempted when using solar filters or solar telescopes specifically designed to observe the Sun safely. Attempting to view the Sun without proper precautions and equipment will result in damage to your eyes.

With the proper precautions, there are a lot of advantages to daytime viewing of the Sun over the nighttime observing of the stars, galaxies, nebulae, and planets:

1. The Sun is dynamic—Solar prominences, sunspots, and other solar activities occur over a matter of minutes. With an H-alpha filter, the great arches of energy that is a solar prominence can be seen as a first blip on the edge of the Sun, develop into great arches, spread outward from the Sun, and dissipate and fade in an hour or two. Sunspots, viewed through a white-light filter, can be tracked across the solar face.
2. The Sun is available every day—So long as it is clear. Partly cloudy skies when the Sun is shining through those little white puffs, can still afford some spectacular views.
3. Light pollution is not a problem—It's daytime. The Sun is bright!
4. Aperture is not king!—The Sun is bright! Aperture provides additional detail, but light gathering is not a concern. Dedicated solar telescopes are available with apertures as small as 35 mm–40 mm and are highly effective for solar observing.
5. Solar telescopes tend to be small in size—Easy to set up. Easily a grab-and-go telescope.
6. Solar telescopes are easy to mount. Small H-alpha telescopes can be mounted on camera tripods. Solar eclipse chasers love the airline transportability of these setups.
7. The Sun is easy to find—A GoTo telescope mount is not needed. Aiming the telescope is a different problem, necessitating a different kind of finderscope than used at night. Remember, no direct viewing of the Sun!

Solar Filters

There are two families of solar filters, white-light filters and Hydrogen-alpha filters.

White-light filters commonly take the form of either a specially coated glass or foil-film mounted in a round frame attached on the front of the telescope. These filters are available in many sizes to accommodate all make and models of telescopes. These are safe, but not indestructible. Handle with care when mounting and dismounting on your telescope. White-light filters enable the observer to use a regular telescope to safely view the Sun and observe sunspots (Fig. 5.18).

Hydrogen-alpha filters are a whole different animal. These exotic pieces of equipment enable the observer to view all sorts of solar phenomena. They can be purchased as a two-piece add-on to your own telescope, with a front of the objective mounted pre-filter and an etalon filter on the eyepiece to perform the filtering down to 1 Å or less. H-alpha (Hα) is a specific deep-red visible spectral line created by hydrogen with a wavelength of 656.28 nm, which occurs when a hydrogen electron falls from its third to second lowest energy level. This is very useful for observing solar prominences. H-alpha filters can be purchased as a stand-alone telescope, with a singular purpose of viewing the Sun safely. Be prepared, viewing the Sun in the hydrogen-alpha light is an expensive proposition. H-alpha setups can run from

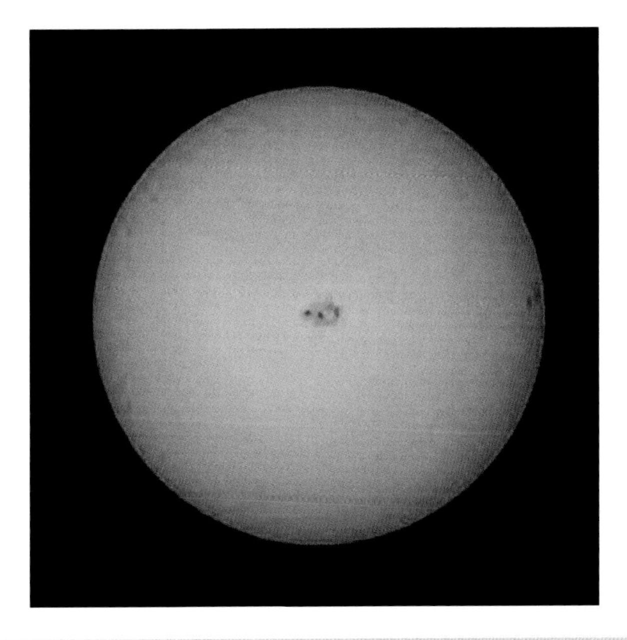

Fig. 5.18 Sunspots on the Sun through a white-light filter (James Chen)

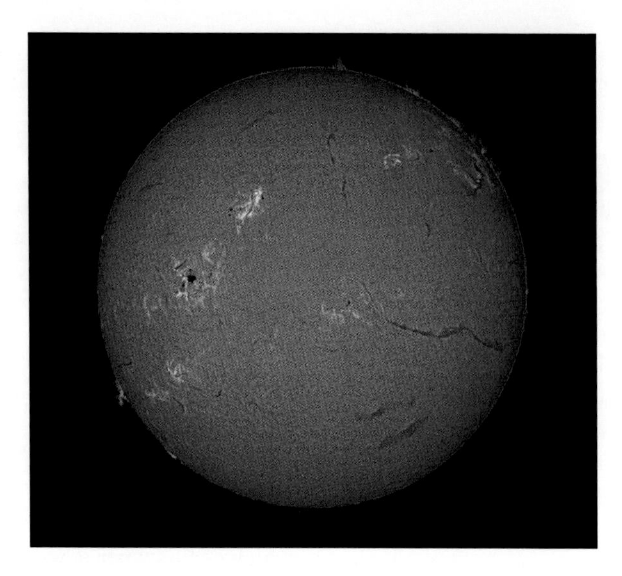

Fig. 5.19 Solar prominence through a hydrogen-alpha filter (James Chen)

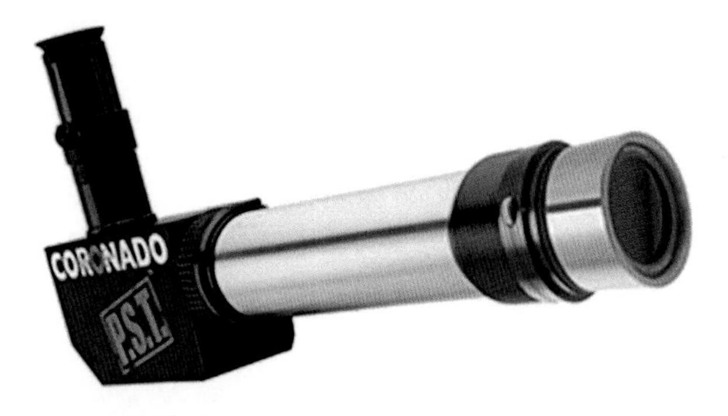

Fig. 5.20 Coronado PST with double stack (Hands-on-Optics archives)

$600 to well over $5000. But the view is spectacular! Most amateurs who own H-alpha equipment say that it is worth the price of admission (Figs. 5.19 and 5.20).

There are cheap solar filters that come as standard equipment with some older telescopes. Do not use these. They attach to the eyepieces and absorb the magnified light and heat from the telescope objective. The heat of the Sun's ray will crack and break these filters in use, resulting in eye damaging solar rays penetrating through the eye and retina. Throw these things away!

There are many supporters of the use of a Herschel wedge to view the Sun in white light. A Herschel wedge is an optical prism used to refract most of the light out of the optical path, allowing safe visual observation, with excellent results.

While the Herschel design uses a prism that is constructed of special glass which absorbs UV and IR light, it cannot be used with a Newtonian reflector or Cassegrain design. This is because reflecting telescopes usually have a secondary mirror close to focus which may be damaged by the heat caused by IR light. This concentrated IR, in particular, can crack or unglue optical elements in a telescope, especially those located close to the focal plane, resulting in serious injury to the eye or possible blindness. The glass objective lens in a refracting telescope never absorbs these wavelengths.

The prism in a Herschel wedge is a trapezoidal cross section. The surface of the prism facing the light acts to reflect a small portion of the incoming light at 90° into the eyepiece. The trapezoidal prism shape refracts the remaining light gathered by the telescope away at an angle. The Herschel wedge reflects about 4.6% of the light that passes through one of the prism faces that is flat to 1/10 of a wave. 95.4% of light and heat goes into the prism and exits through the other face and out the backdoor of the housing, thus the excess light and heat is dispersed and not used for observing. It is also important to note that the Herschel wedge will still allow 4.6% of the light from the Sun to pass through, which is still strong enough to burn the retina, and so an appropriate neutral density filter must still be used. Without the additional filtering, eye damage will result. Herschel wedges are expensive, can only be used with refractors, and without additional filtering are hazardous used on their own. Buyer beware.

For binocular users, there are white-light solar filters available for mounting on the front of binoculars. One major solar telescope manufacturer in the past sold dedicated solar binoculars. While very convenient to carry around, the use of solar filtered binoculars takes a little skill to use. With the majority of light filtered out, it is actually difficult to locate the Sun through solar filter binoculars. A view of blackness is present until the image of the Sun suddenly pops into view.

All the discussion here has been on white-light and Hydrogen-alpha solar observing. There are solar telescopes and filters that enable the observer to view the Sun in the Calcium K lines. Calcium K (Ca-K) telescopes and filters are used to study the wavelength of 393.4 nm. This emission line is one of two that are produced by calcium just at the edge of the visible spectrum, in a layer that is slightly lower and cooler than the layer viewed in Hydrogen-alpha. The emission line displays areas of super granulation cells that are brightest and strongest in areas of high magnetic fields, such as sunspot activity and active regions.

While scientifically important, and photographically accessible by older amateurs, this is a difficult solar image to observe visually. There is an age-related component to visual viewing through a CaK telescope or filter. Teenagers are reportedly more apt to see details in the dim faint bluish images, and those over 40 mostly claim to see nothing at all. The recommendation here is to avoid the CaK solar telescopes, as the likelihood of a 60+ year old observer seeing anything satisfying is quite remote.

Recommendations for Older Astronomers

Compromise. That is the key word when selecting telescope gear as you get older. The trade-off is between aperture and size versus portability and ease of use.

Every amateur astronomer goes through a period of their life in this hobby where aperture fever takes over. For those new to the hobby, aperture fever is the never-ending desire for a larger (as in diameter) telescope to gather the valuable photons emitted by distant galaxies and stars. The mantra of many astronomers is "Aperture Rules."

This can be seen in the professional astronomy ranks. For years, the largest telescope in the world was the 200″ Hale telescope at Mt. Palomar. For those comfortable with the metric system, the Hale has a 5.08 m aperture. For those born in the 1950s and 1960s, the Hale telescope was legendary. The 1975 Russian-built 6 m failed to supersede the ultimate status of the Hale (Fig. 5.21).

Then, beginning in 1993 with the building of the 10 m Keck telescope in Hawaii, 18 ground-based observatories housing telescopes of 6–10.4 m have been built in astronomy's quest to view the deepest reaches of space. Aperture fever on a grand scale (Fig. 5.22).

Even in space-based telescopes, aperture fever is evident. Even the historically significant 2.4 m Hubble Space Telescope will be replaced by the larger James Webb Space telescope with its 6.5 m mirror array (Figs. 5.23 and 5.24).

Fig. 5.21 The 200-in. Hale telescope at Mt. Palomar (Caltech.edu)

Fig. 5.22 The Keck 1 telescope on Mt. Mauna Kea, HI (Roger Ressmeyer/Corbis)

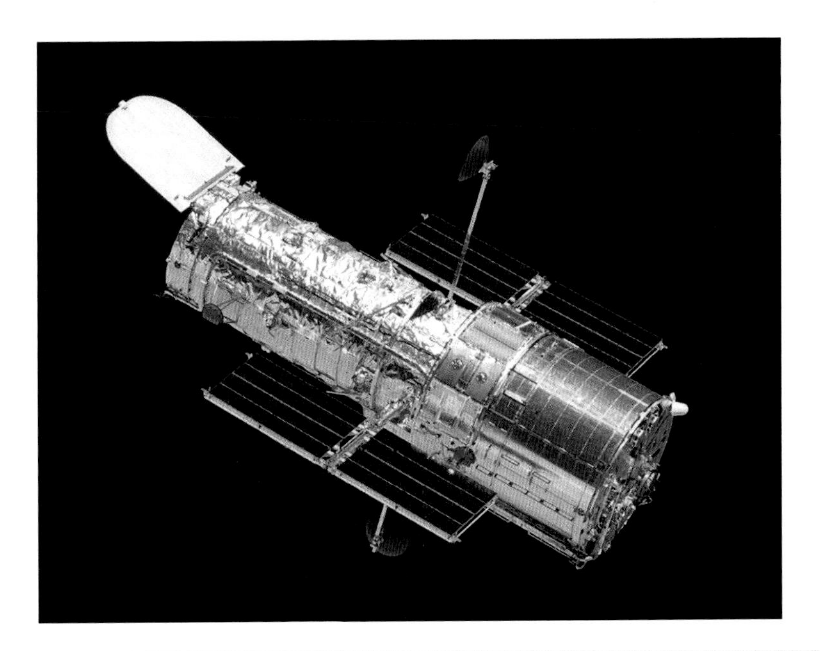

Fig. 5.23 The Hubble Space Telescope (NASA)

Revisiting our friend John, who has for years used a Celestron C14, it's time for him to obtain a smaller telescope that he'll be able to use on a daily basis. He's getting older and despite good health and fitness, he's not getting any stronger with age. The recommendation for John is to get a small 6-in. or 8-in. SCT. Although

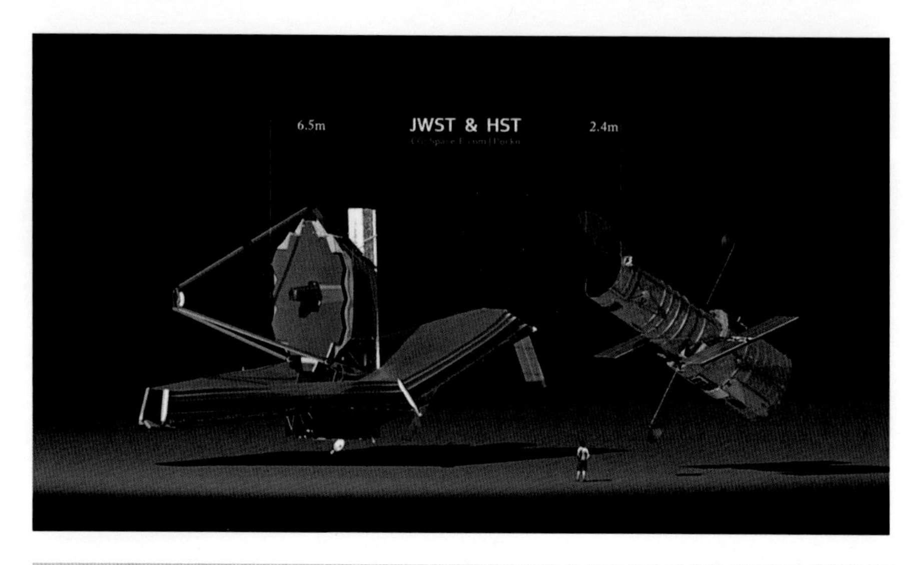

Fig. 5.24 Artist's comparison of the James Webb Space Telescope versus the Hubble Space Telescope (NASA)

smaller in aperture, this size telescope will receive greater use. And he can still keep the big 14-in. for the once per month special observing time.

And alternative is to acquire a high-quality 102 mm–120 mm apochromatic refractor. The mounting requirements for this size refractor are not overwhelming and still easily portable for a 70+ year old man.

A comment should be made at this juncture. There is an anecdotal story of a 70-year-old man who was shopping for a cutting edge quality apochromatic telescope in the Washington, D.C. area about a decade ago. This gentleman could have obtained a premium quality 102 mm Japanese-made apochromatic refractor and received delivery within 2 weeks of placing his order. Instead, he chose to place an order for a similar size telescope with a well-known premium quality American refractor manufacturer that had a waiting list of 3 years. Under normal circumstances, this would not be an unreasonable decision. But at age 70, with the delivery of the ordered telescope when he is at age 73, this decision is questionable. Price was not an issue. A performance comparison between the two telescopes revealed them to be arguably equivalent. The question arises: Why wait 3 years? With declining eyesight, waning physical strength, and other vulgarities of growing old. It's closing time and last call for alcohol! We're not getting any younger!

Alternatively, mounting a medium or medium-large telescope on wheels is a viable option for older backyard astronomers. These wheeled accessories can be homemade do-it-yourself items, or can be purchased commercially such as the JMI Wheeley Bars or ScopeBuggy. Some pre-planning is required for their use. Telescopes can be moved by an older person quite easily from inside a home, storage shed, or garage onto a driveway, patio, or deck. Considerations for these

Fig. 5.25 A Dobsonian telescope mounted on JMI Wheeley Bar (Jim's Mobile Inc.)

wheeled accessories are finding a level area to park the telescope in the storage area and eventual observing area in the yard, and the smooth transition and clearance of the telescope through doors (Figs. 5.25 and 5.26).

For married male readers, this is probably a good time to discuss the concept of WAF—Wife Acceptance Factor. Some telescopes are very attractive in a living room or family room environment and can add to the décor as a point-of-interest. These attractive telescopes have a very high WAF rating. However, some telescopes are industrial-looking and ugly, and have a low WAF rating. The addition of wheeley bars only adds to an industrial look that can only be acceptable in a home décor based on a warehouse or the Steampunk look.

For many reasons, as the backyard astronomer gets older, the storage of the wheeled telescope in a car garage or weather-protected outdoor shed is recommended.

The ultimate telescope accessory is a backyard observatory. Whether from a kit, building plan, or self-designed and built, these structures are the ultimate for an amateur astronomer, young or old (Fig. 5.27).

The major considerations for an observatory are providing shelter for the telescope and keeping it from moisture. Humidity and moisture are the deadly enemies

Fig. 5.26 Tripod-mounted JMI Wheeley Bar (Jim's Mobile Inc.)

Fig. 5.27 DIY observatory (James Chen)

of telescopes, telescope optics and eyepieces, telescope mounts, and mount electronics. The goal of any observatory is to have the telescope and its optics be at ambient temperature and DRY! That means a building that is airtight and waterproof. Moisture can corrode mechanical parts, reflective optics coatings can deteriorate, fungus can grow on lenses, and mount and camera electronics can corrode causing short or open circuits.

As one grows older, the maintenance of anything, home maintenance, automobile maintenance, and in this case telescope and observatory maintenance becomes an issue. There is a true story of a 92 amateur astronomer who had passed away. His observatory was neglected by him for years, and as his widow brought in experts to handle the disposition of the contents of the observatory, it was discovered that the building had developed a leak in its dome. The classic Cave 12.5-in. convertible Newtonian/Cassegrain telescope and its mount had rusted and corroded, the optics coating had virtually disappeared, and the electrical wiring had dried and cracked. What seemed to be an ideal solution for the telescope had become a useless collection of telescope parts because of the moisture and the lack of maintenance.

Real estate is also a consideration if the observatory route is to be followed. A simple observatory, even when it's a DIY project, will likely cost $5000 or more. An observatory should not be located in an urban or heavy suburban environment, because of too much light pollution and buildings and other houses obstructing the view. A country location away from city lights is perfect, but not necessarily convenient or close to other amenities such as shopping, cultural centers, movie theaters, etc.

So what is the best path to follow? Smaller and easier to set up and operate telescope? Wheeley bars equipped telescope and storage in the garage? Or an observatory?

The answer is: Yes!

The answer is it depends on the type of person you are. The majority of people will probably find the smaller and very portable high-performance telescope as the perfect choice. For those amateurs who grew up with large telescopes and don't want to give them up, the wheeley bar and observatory approach are good options. If a lifetime of meticulous maintenance is in your background, the observatory is an excellent, if expensive, option. Otherwise, the wheeley bar route is an attractive option.

The strategy for a first-time older astronomer is to buy a quality smaller and portable telescope, such as an 80–102 mm refractor on either an alt-az or equatorial mount. Or a 6–8-in. SCT on a GoTo mount. Or a 10-in. Dobsonian. Any of these telescopes can be used as the "grab-and-go" telescope.

The strategy for experienced amateurs who already owe a big telescope, buy a quality smaller and portable telescope, such as a 80–102 mm refractor on either an alt-az or equatorial mount. Or a 6–8-in. SCT on a GoTo mount. Or a 10-in. Dobsonian.

And for the Charlene's of the world with their observatories, but no portable capability, buy a quality smaller and portable telescope, such as a 80–102 mm refractor on either an alt-az or equatorial mount. Or a 6–8-in. SCT on a GoTo mount. Or a 10-in. Dobsonian.

There seems to be a pattern in these recommendations …

Stay Seated

Most telescope tripods come with extendible legs.

At many star parties, telescope tripods are extended to allow people of all stature to access the telescope's eyepiece.

This does not mean that during private sessions, the older backyard astronomer needs to go to this extreme.

There are numerous height adjustable chairs on the astronomy market for use while observing. Some are just adjustable musician or drummer's stools. Some are designed to use gravity and friction and the observer's weight to lock in the height of the seat. And some are like Fig. 5.29, where more mechanically solid horizontal bars anchor the observer's seat to the desirable height. Try before you buy. The drummer stool doesn't have the range of heights needed by some users. The gravity–friction models sometimes slide down at inopportune moments. The illustrated observer's chair has been used to great success (Fig. 5.28).

Fig. 5.28 Star dust chair (Hands-on-Optics)

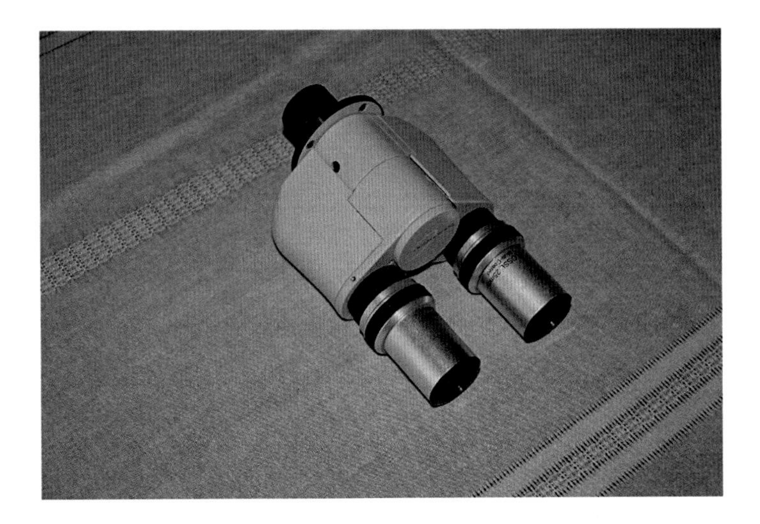

Fig. 5.29 The author's Denkmeier binoviewer (James Chen)

Binoviewers

"Two eyes are better than one," the old saying goes, but telescopes aren't exactly designed with this in mind. If you've endured eyestrain and pesky "floaters" in your eyes ruining your view of Jupiter or the Moon, you understand the drawbacks of "cyclops" astronomy. But there's a solution: binoviewers. Few other accessories can add horsepower to your telescope like a pair of binoviewers, especially for observing the Moon, planets, and smaller celestial objects (Fig. 5.29).

Here's what you need to know about these powerful observational tools:

1. Binoviewers are a packaged set of prisms that split into two the light path from your telescope and direct each beam to a separate eyepiece. There are two main types available. The less expensive is based on designs for lab microscopes. They tend to use smaller prisms which can vignette (cut off) light from some lower-power eyepieces, which are just the ones you want to use to get wide-field views of star fields. They work quite well at moderate magnification.
2. The more expensive binoviewers use larger prisms and accommodate wide-field, low-power eyepieces, so they give crisper and more expansive views than less expensive optics. They also hold focus as you adjust the interpupillary distance. But they are heavier and much more expensive … at least $1000–$1500 (not including eyepieces). These are a good choice if you have an excellent telescope, a solid mount, and a budget that can handle it.
3. Of course, both types of binoviewers require two identical eyepieces, so this also adds to your expense. Eyepiece collections get quite large and expensive very quickly when using binoviewers.

Fig. 5.30 The author's Stellarvue SV80BV/Denkmeier binoviewer setup (James Chen)

The main drawback to binoviewers is a less bright image. Since they split the light into two beams, binoviewers reduce image brightness to each eye by at least a factor of two. But the brain recombines the two images to produce a perceived total brightness nearly equal to that of a single eyepiece telescope. The effect is to make an image of a 10-in. scope to appear as bright as a single beam from an 8-in. scope, for example. Most binoviewer users report about a 0.5 magnitude drop in light gathering when compared to single eyepiece observing. But the effect is not as pronounced, because the brain processes the images from both eyes. The loss is not noticeable on bright objects, such as the Moon, Jupiter, Saturn, or Mars at opposition. Galaxies and nebulae right at the limit of your telescope will be dimmed significantly (Fig. 5.30).

The trade-off for the slight light loss for binoviewers is by using both eyes, the brain has two images to process as it is normally used to, with greater detail as a result. To use an electronic term, the image that the brain is able to process has a greater signal-to-noise ratio. Floaters in the eyes become less noticeable.

There is another caveat to binoviewer use. Binoviewers add another 4–5 in. of light path, so the focuser of a telescope needs enough travel to bring the image to a focus. This is usually not a problem with Schmidt–Cassegrain scopes, which you focus by moving the primary mirror itself. Some refractors have been marketed as binoviewer-ready by having a shorter tube. These binoviewer-ready telescopes then supply extension tubes that insert into the focusers to allow single eyepiece viewing. But many refractors and nearly all Newtonians do not have enough travel in their focusers, unless you add a Barlow lens to bring the image to focus. And a Barlow lens increases the magnification of each eyepiece by 1.5×–2×, which means a smaller field of view.

Are binoviewers worth the added expense?

For older observers, especially those bothered by eye fatigue and floaters in their eyes, binoviewers are a solution. Since both eyes are being used, the brain processes out the floaters and the eyes are less fatigued.

Looking at the Moon and bright planets through a scope equipped with binoviewers is a stirring experience. The Moon takes on an almost "3D" appearance like you're looking out of the window of a spaceship. The bands of Jupiter and rings of Saturn snap into view clearly. And looking with two eyes seems to reduce eye fatigue and the effects of distracting "floaters" that move in and out of view.

A favorite viewing pleasure of this author is to use a pair of 5 mm eyepieces in the binoviewer when viewing the Moon through a 102 mm refractor. By turning off the clock drive of the telescope, and allowing the Moon to drift by, the view through the binoviewer is like being in the Apollo command module in orbit and looking down upon the lunar surface as it speeds by Breathtaking!

Chapter 6

Astronomy Clubs, Public Outreach, Star Parties, and Staying Social in Later Years

Star light, star bright,
the first star I see tonight,
I wish I may,
I wish I might,
Oh darn,
it's just a satellite!

—anonymous

There are a number of activities and organizations that can keep an active senior-aged astronomer socially connected and interactive in the community. First and foremost is the local astronomy club, where seniors can meet people of all ages from their community and nearby areas who share the same interest in astronomy. Most astronomy clubs support public outreach and provide their school districts with science education opportunities.

Ultimately, every amateur astronomer makes their way to a star party, whether sponsored by their local astronomy club or traveling to a remote dark site to attend a regional or national star party event. Many plan their year around the large nationally recognized star parties.

J.L. Chen, *Astronomy for Older Eyes*, The Patrick Moore Practical Astronomy Series, DOI 10.1007/978-3-319-52413-9_6

Astronomy Clubs

There are astronomy clubs in every state of these United States and in many countries of the world. Everywhere people own and use telescopes to view the night sky, there is an astronomy club available for astro-minded people to join.

An astronomy club is a great opportunity for older amateurs to socialize and stay connected.

Astronomy clubs come in various flavors. Most are generalized, trying to address all interests in astronomy. But some clubs are specialized, and most clubs will have a particular strength in specific goals or activities:

1. Public education and outreach
2. Telescope making
3. Astrophotography and astro-imaging
4. Classic telescopes and telescope restoration
5. Computers and telescope automation
6. Planetary observations
7. Deep Sky observing
8. Satellite tracking
9. Meteor observing and meteorite hunting
10. Radio astronomy
11. Just plain socializing

Astronomy enthusiasts, as a community, are very welcoming and eager to exchange thoughts and ideas, and share scientific knowledge. Joining an astronomy club is an excellent way for an older individual to stay connected with people, exercise their communication skills, enhance their knowledge base, and stay current on the latest science and technology.

Public Outreach

The goal of many astronomy clubs is to stimulate greater public interest in astronomy and to assist everyone in becoming more engaged in activities that allow them to learn more about the universe.

In many areas of the country, astronomy and science education in schools is an uphill fight. For example, in 2015, a West Virginia high school principal deemed a school planetarium as unnecessary, trashed all the astronomy books and shipped to the dump an expensive planetarium projector in order to gain room for the high school football team training space. This was a very short-sighted, sad, and disappointing outcome for a school and its students. The school administration and its educators should know better.

In order to encourage students and their parents to pursue astronomy and scientific knowledge, astronomy clubs sponsor educational activities that include:

1. Promoting observational astronomy nights in cities and small towns. Astronomy clubs will coordinate star party nights to give many people their first look at the Moon, Saturn, and double stars. The astronomy clubs also will provide astronomy talks and helpful literature that will excite and encourage newcomers about entering the astronomy hobby.
2. Coordinating outreach efforts with astronomy groups and star parties. The clubs will help with organization and publicity for astronomy get-togethers and group functions by providing materials and guidance to schools, summer programs, and other groups and activities.
3. Increasing the presence of astronomy in schools. Even teachers interested in astronomy in secondary schools have little time to devote to it. Astronomy clubs provide lectures and talks about astronomy, observing, and how to use telescopes and other equipment to engage and enliven current science curricula. Many members of astronomy clubs are themselves teachers and educators.
4. Interconnecting the pure sciences of physics, chemistry, and mathematics with the science of astronomy. For example, Einstein's Theory of General Relativity has real relevance in astronomy and the origin of the universe. Another example is how the spectra of star provides information to scientists of the chemical makeup of stars. Additionally, how mathematics provides a descriptive model of the workings of the universe.
5. Spreading interest in astronomy through social media. A new generation of astronomy enthusiasts will ultimately come largely through outlets such as Facebook and Twitter, where Generations X and Y spend much of their time. Astronomy clubs actively pursue a presence and activities on social media outlets.

Star Parties

For many long-standing amateur astronomers, a star party is an opportunity to socialize with like-minded people in the love of their hobby. Showing off their telescopes, eyepieces, and demonstrating their capabilities is also part of the star party experience.

For those new to the hobby, a star party is an opportunity to look through telescopes, see things they have never seen before, meet experienced astronomers, ask questions, check out equipment, and experience the wonders of the night sky. Not bad for a weekend activity.

So what is a star party? The Wikipedia definition is: "A star party is a gathering of amateur astronomers for the purpose of observing the sky". Local star parties may be one night affairs, but larger events can last up to a week or longer and attract hundreds or even thousands of participants."

Many astronomy clubs, state parks, and National Parks sponsor public outreach and single night observing sessions, where individuals set up their telescopes at a park and invite the public to gaze at nighttime sky objects such as planets, galaxies, and nebulae. Often, a club member or a park ranger will provide a talk about astronomy and describe what the attendees will see prior to an evening's viewing session. These local outreach events will often coincide with an astronomical event, such as a lunar eclipse, meteor shower, or Mars opposition.

Big event star parties are an altogether different affair. These are annually held, multiple day events, often beginning on a Friday and ending on Sunday or Monday morning. Some of the large star parties are held over a week. Larger star parties feature lectures, swap meets, displays of carefully handcrafted DIY telescopes, contests, tours, raffles, and other similar activities. Commercial vendors are often invited to sell a variety of astronomical equipment, allowing the attendees the opportunity to "try before they buy." There is much meeting-and-greeting and discussion of various aspects of the hobby at every star party.

Typical Large Star Party

The blueprint for the big event star party was established by the grand daddy of all star parties, the annual Stellafane convention held on Breezy Hill in Springfield, Vermont each year. Stellafane began in 1926 with a gathering of 20 amateur telescope makers, and has now grown to an annual attendance of over 4000. Held during a new Moon weekend in late July or early August (it varies every year), Stellafane draws people from all walks of life to share their love of astronomy every year. (Figs. 6.1 and 6.2)

Many big event star parties, such as RTMC (formerly Riverside Telescope Makers Convention) near Big Bear Lake in California, the week-long Texas Star Party at the Prude Ranch (Fig. 6.3) near Fort Davis, Texas, and the Winter Star Party (Figs. 6.4, 6.5, 6.6, and 6.7) in the Florida Keys, all follow the Stellafane blueprint:

1. Gates to the star party open mid-day on Friday. Attendees have already pre-paid their entrance fee, or pay at the gate a slightly higher fee. There is often a long line of cars parked by the side of the road leading to the main gate of the star party awaiting the opening of each event.
2. Attendees set up their camp site and telescopes during the Friday afternoon. Many attendees may choose to stay at a local hotel and enter and exit the star party at will. Some star parties are held at boy's club grounds or boys or girls scout camp grounds with bunkhouse sleeping facilities. Camp grounds may or may not have bathroom facilities, although porta-potties abound. Be prepared for some rather crude facilities. For years, the shower stall that was available at Stellafane consisted of a 3-sided black plastic wood frame platform and a garden hose. Privacy was definitely lacking. The water was COLD! The spray nozzle was not easy to use with soapy hands. Stellafane has rectified this with somewhat less-crude facilities.

Fig. 6.1 The pink Stellafane Clubhouse during the 2001 Stellafane Star Party (Stellafane)

Fig. 6.2 Looking at the telescope field down Breezy Hill at Stellafane (Stellafane)

Fig. 6.3 Telescope setup at the Texas Star Party (Hands-on-Optics archive)

Fig. 6.4 The Winter Star Party (James Chen)

3. Friday afternoon talks are given by astronomy writers, scientists, and telescope makers. Often, scientists and astronomers from NASA or from a major observatory are invited to present their latest discoveries, or a major telescope manufacturer will introduce new products.

Fig. 6.5 The Winter Star Party (James Chen)

Fig. 6.6 Winter Star Party cont'd (James Chen)

Fig. 6.7 Winter Star Party cont'd (James Chen)

4. Many star parties will conduct a grind-your-own-mirror course or build-your-own-telescope class, enabling attendees to grind their own Newtonian mirror and be able to build their own telescope. Many of the star party events owe their existence to amateur telescope makers. Stellafane began as an amateur telescope makers convention. RTMC stands for the Riverside Telescope Makers Convention. There are still DIY telescope makers, and these star parties allow them to show off their craftsmanship and handiwork. The telescope making clinics allow this craft to carry on, and affords the opportunity for older astronomers to explore a new hobby, learn a new skill, and build their own telescope.
5. A Friday evening welcome session is held, followed by astronomy talks. Observing discussions for the evening viewing are often the subject.
6. Weather permitting, Friday evening observing starts as soon as it gets dark. Don't be surprised to find people observing throughout the night!
7. Early in the morning, while many late night observers are still sleeping, others start setting up and begin selling at a swap meet. Hundreds of telescopes, eyepieces, filters, mounts, and any astronomy-related paraphernalia can be bought, sold, and traded at these telescope shark-infested feeding frenzies. Bring cash and be prepared to bargain!
8. Saturday afternoon talks are given by astronomy writers, scientists, or telescope makers.

9. Don't forget to enter the raffles at these star parties. Star parties are fund raisers for the sponsoring astronomy club. Vendors contribute astronomy merchandise to the astronomy club for door prizes and raffle jackpots. Astronomy books, star atlases, binoculars, eyepieces and diagonals, and telescopes are often raffle prizes.
10. Saturday evening usually features a large gathering, with some form of dinner meal, noted guest speakers, and the raffle drawing for prizes.
11. Following the main meeting, everyone ventures forth to the observing field to engage in a night's worth of observing. This is the true highlight of these events. For many, a star party is the only time people can escape from their light polluted urban and suburban neighborhoods and really experience dark skies.
12. Sunday morning is quiet time. Many stayed up throughout the night observing, and went to bed at dawn.
13. By late Sunday morning, people have packed their tents, packed up their telescope gear, loaded their SUV and have left the star party homeward bound.
14. The host astronomy club cleans up the grounds and starts planning for the next year.

Star Party Etiquette

Whether a veteran of star parties or attending for the first time, it is necessary to understand the ground rules of conduct at a star party. A star party is not a raucous party as the name might imply, but a quiet, cerebral and spiritual gathering for the astronomy and science oriented.

The list of observing field rules may seem long and detailed, but largely these rules center around two things, preserving the dark adapted eyes of all observers on the observing field and respect for the telescope equipment on the field. Deep sky objects are very dim, and astronomer's eyes must be dark adapted in order to even see many of them. Dark adaptation takes approximately 30 min in the absence of light. One flash with a white flashlight can ruin the next 30 min of observing. Experienced observers understand this. Beginners or those newly interested in astronomy may not. No white light flashlights are allowed on the observing field. Low power red light flashlights are fine as long as they are directed downward and not flashed in the eyes of observers.

Telescopes on the observing field may range in price from a hand-built hundred dollar Dobsonian to a multi-thousand dollar showpiece telescope. These are precision observing instruments and have delicate parts. Telescopes are prized possessions of their owners. The scope owners have a right to have their equipment protected and respected. Do not touch or lean on any scope without the permission of the owner. Parents are responsible for their own children. Parents should supervise them closely as they may not appreciate the value of these instruments. Most amateurs are more than happy to share a view with you and guide you through the observing experience, and only ask that you respect their equipment in return.

This is by no means a requirement, but feel free to bring a telescope. Don't be embarrassed if you aren't familiar with it or think it may be too small. You will find help using it and may be surprised at what you can see with it. Binoculars are another great way to observe the night sky, so if you can't bring your scope, then bring your binoculars.

Be watchful, guarded, and show respect for all attendees and their equipment. It rarely happens, but very occasionally an eyepiece or telescope accessory goes missing on the observing field. Sometimes it's an accidental drop into the grass to be found the next morning in the daylight. As a whole, amateur astronomers are law abiding and respectful of each others gear.

But when dealing with the public, be on guard. Sometimes an attendee may have "sticky" fingers. This author once had a pair of 18 mm Abbe orthoscopic eyepieces from a binoviewer disappear at a star party, never to be seen again. Not cool!

Keep an eye on your equipment. Make sure eyepieces are stored in a convenient and guarded way that prevents sticky fingers.

Car lights can be a problem at star parties. Do not shine car lights on the observing field. Some newer vehicles have automatic headlights that come on when the vehicle is started, and this can be a problem. Some vehicles have the ability to turn this feature off, others do not. Sometimes partially depressing the emergency brake (but not enough to actually engage it) will disable the automatic lights. Another solution is to cover the headlights.

Parking at many star parties is usually established far from the observing field, and with finesse the headlights are pointed away from the observing field. If lighting up the area is inevitable, warn observers to cover their eyes before illuminating the observing field.

Flashlights, especially red LED lights, are essential at star parties in order to navigate around the observing field. Once the eyes are dark adapted, many find that they do not need a flashlight. Do not use a bright white light flashlight. Bring a dimly lit red light flashlight. Many vendors at the Saturday morning swap meet will be selling red LED flashlights. Avoid shining the light in an observer's eyes.

No flash photography, period. There are no exceptions. Don't spoil everyone's dark adaptation. Don't spoil an astrophotographer's extended exposure.

Be aware that some attendees will be engaged in astrophotography. There will be attendees who are avid deep sky imagers. If they are in the middle of taking an astro-image, don't disturb them. Feel free to bring a tripod and camera with a big lens and try some long exposure photography. Star parties are an excellent opportunity to seek guidance for first-time astrophotographers and astro-imagers. There will be dozens of experienced amateurs who can help beginners produce their first images. But do so on a noninterference basis.

Avoid use of green light laser pointers during astrophotography sessions. These devices can ruin an evening's imaging efforts.

No smoking. It's not allowed, no exceptions. Smoke can be harmful to telescope optics and they are a bear to clean. Just don't smoke around telescopes. P.S. People don't like smoking too!

Check with the star party organizers on the subject of alcohol. Alcohol may be prohibited. Some beer or wine may be allowed depending on local rules and ordinances. No hard liqueur, a star party is not New Year's Eve or Spring Break! Again, check with the star party sponsor or the park ranger.

No aerosol sprays on the observing field. One drop of errant insect repellant or suntan lotion can permanently damage the delicate optics or coatings of a telescope or binoculars. If you apply insect repellant, do so a significant distance away from the telescope field, and be aware of the wind direction. Don't allow the insect repellant application to drift towards the telescope field in the wind. Carefully and thoroughly wash your hands afterward.

Some observers enjoy playing music while looking through their scopes. This is acceptable under certain circumstances. A traditional and acceptable piece of music is Holst's *The Planets*. Please have respect for those nearby and ask before playing music. If agreed to, keep the volume to a reasonably low level in the observing area. Not everyone enjoys music while observing. Also, please select music that is pleasing to most everyone. No rap, hip-hop, or heavy metal music. When in doubt, use earbuds and an iPod.

Take care and clean up all the trash around the observing site. The common sense rules apply here. Make a mess, clean it up. Bring trash bags to place trash in. If available, place recyclables in proper recycling bins. Many star parties are held in public campgrounds, state or national parks. Not only can trash be unsightly, but in some parks, trash can attract wild animals, such as bears and coyotes. A hungry bear can do considerable damage to an astronomer or the astronomer's telescope. Help keep the facility looking nice and presentable. Ultimately all attendees of star parties are all guests at the site. Don't wear out the welcome.

All star parties observe quiet mornings. Most observers come to a star party simply to observe from dusk till dawn. All night observers will go to bed as the sun is rising. If you are an early riser, please have respect for those still in their tents or sleeping bags. Please try to be as quiet as possible at least until the majority are up and around.

Children are the future of amateur astronomy and parents are encouraged to bring them along to the party. A star party can be a very exciting time for most everyone, kids included. Keep an eye on the children. There are tens of thousands to millions of dollars in equipment out on the observing field, depending on the size of the star party. Most scope owners have saved for years to buy their dream scope, or have countless hours in building their own equipment. Children should be instructed not to run or play around the equipment on the observing field. They should also get the owners permission before touching any equipment.

At some star parties, an event official will issue only one warning if your child or children are breaking the basic rules. After that, parents and their troublesome children may be asked to leave the premises. It is dark, accidents can happen with unruly children, and nobody wants any injuries or damaged telescope equipment. The goal is to maintain optimal conditions for observing and keep equipment free from harm.

Do not touch a telescope unless the owner has given permission. Never touch the optical glass of a telescope or eyepiece. The skin oils can ruin the coatings. Sometimes delicate alignments can be accidentally altered.

There will be some really big telescopes at most star parties. While observing at night at a star party, there will be lines of people forming behind certain telescopes, waiting for a chance to view through a special optical instrument. Many telescopes are short enough that one can simply walk up and look in the eyepiece, or better yet, can sit down and look through it. SCTs and refractors come to mind. Then there are the big Dobsonians. 25-in. or 30-in. Dobsonians will have eyepieces located where a 10 ft ladder is required to reach the eyepiece. It is not uncommon to climb 4, 5, even 6 ft up a ladder just to look through the eyepiece. A 30-in. *f*/6 pointed at the zenith will require an observer to climb a ladder to get their eye level to the 7-1/2 ft or 8 ft level! No big deal you say? They can be! Just remember, you are doing this in the dark. When you do go up a tall ladder, be sure to count your steps. If you forget that you are on a ladder and turn around to walk away, that first step could mean a trip to the hospital. If in doubt, ask the owner to count you down. It's better to be safe than sorry. And whatever you do, DON'T TRY TO BREAK YOUR FALL ON THE TELESCOPE!

Scope owners will not be held responsible for injuries to attendees climbing their ladders. But attendees will be held responsible for any damage caused by carelessness to a scope. By climbing the ladder, each person assumes the responsibility of getting up and down safely. For older people who know that they can't get up and back down a ladder safely, don't go.

Before night falls on the first night of the star party, stroll around the grounds (until you feel at home!).

It's a lot of fun to walk around the observing field during daylight hours, checking out all of the beautiful telescopes that are set up. Admire the craftsmanship of homemade telescopes and the excellence of high-quality commercially made telescopes can be equally admired. A daytime walk around the grounds also familiarizes everyone with the locations of everything in the dark. Take note of where the bathrooms or porta-potties are located. Take note of telescope mount legs and power cables for future reference in the dark. At night, use a red filtered flashlight while walking about the field and keep it pointed towards the ground. Be on the lookout for power cords and tripod legs. Once eyes become dark-adapted, walking around the field in the dark becomes easier.

Be dressed appropriately. Give some thought to this one! Check the local newspapers or the Weather Channel prior to an observing session. Astronomy is not much fun under uncomfortable conditions. Standing around in the night air can be quite cold, even in the summertime. Bring a long-sleeve shirt or a sweater. Long pants and a long sleeve shirt also help deter bugs. During the winter, dress warmer than normal. While standing around looking through a telescope on a cold winter's night, the human body generates no additional heat from physical motion, The cold will seem colder.

Bring insect repellent and lots of it. Be willing to share it. Remember the outdoor nature of this hobby, and since the majority of star parties are held during the summer and early autumn months, be aware that some locales have mosquitoes and other biting insects. Take precautions and bring insect repellant. Again, avoid aerosols and apply the repellant away from telescopes, eyepieces, and other telescope accessories. Remember, outdoor camping stores also sell clothing impregnated with insect repellant!

Don't be afraid to ask questions. Everyone attending a star party is there because of their love of the hobby and the science. The only bad question is the one that isn't asked. A star party is the best classroom for learning about the hobby. Whereelse can one find such a concentration of amateur astronomers and telescope makers! The lessons learned during the 2 or 3 days of attending a star party are invaluable. The knowledge and information gained by asking questions and listening to discussions will be more than can be accrued in a year out on your own.

Fun, Fun, Fun is the operative words for star parties. The enjoyment of a star party is indescribable. Many new friends will be made, the views through many different telescopes will be astonishing, and a feeling of wonder, satisfaction, yet wanting more will be pervasive. Most attendees are already looking forward to the next star party even before they leave the observing field.

The sponsors and club members of each star party make every effort for an enjoyable star party. If there are any comments, questions, or concerns at the star party, feel free to let one of the club members know. The end result will be a better event next year.

Chapter 7

Physical and Environmental Challenges of Astronomy in Later Years

I was lying on my bed staring at the stars,
then I asked myself:
"Where the heck is the ceiling?"

—Instagram joke

For backyard astronomers, the warm humidity of summer observing gives way to the cooling air of autumn. The celestial objects of summer also give way to clusters, nebulae, and galaxies of the fall. The cool brisk autumn then gives way to the cold and darkness of winter, with a whole new set of deep sky objects to observe. Cold winter then yields to spring, with galaxies to explore. Then a return to the warmth of summer. The Earth's position in its orbit around the Sun also changes the perspective of the sky. The summer has astronomers gazing into the heart of the Milky Way, while in the colder months the attention is more towards intergalactic treats.

As the seasons change, the type of clothing changes, especially for older amateur astronomers. Older astronomers must be extra attentive to dressing for the climate and environment. The older one gets, the less tolerant of temperature changes and humidity one is.

The Threat of Hypothermia

Older observers have learned that the cold weather feels colder than it felt at a younger age. This is a major concern, because the older body responds to the colder weather differently than at a younger age.

J.L. Chen, *Astronomy for Older Eyes*, The Patrick Moore Practical Astronomy Series, DOI 10.1007/978-3-319-52413-9_7

Older adults can lose body heat faster. A big chill can turn into a dangerous problem before an older person even knows what's happening. This is called hypothermia. Hypothermia happens when your body temperature gets very low. For an older person, a body temperature colder than 95° can cause many health problems such as a heart attack, kidney damage, liver damage, or worse. The aging astronomer must use a strategy for preparing for an observing night in a warmer fashion than a younger self would.

As the backyard astronomer experiences the cooling weather, some lessons of physical preparedness are required.

1. Never underestimate the cooling temperatures. Since observing through a telescope means sitting quietly with eye to eyepiece, the chill of the night can have an adverse effect. Dress in layers and warmer than necessary. Wear several layers of loose clothing when it's cold. The layers will trap warm air between them. Don't wear tight clothing because it can keep your blood from flowing freely. This can lead to loss of body heat. Coats, hats, gloves, and blankets are the dress code. Hunting and camping stores are a good source for cold weather gear, such as heated socks and heated gloves and warming packs. (Figs. 7.1 and 7.2)
2. Ask your doctor or pharmacist how the medicines you are taking affect body heat. Some medicines used by older people can increase the risk of accidental hypothermia. These include drugs used to treat anxiety, depression, or nausea. Some over-the-counter cold remedies can also cause problems.

Fig. 7.1 Celestron Firecell Hand Warmer/Power Packs (Celestron)

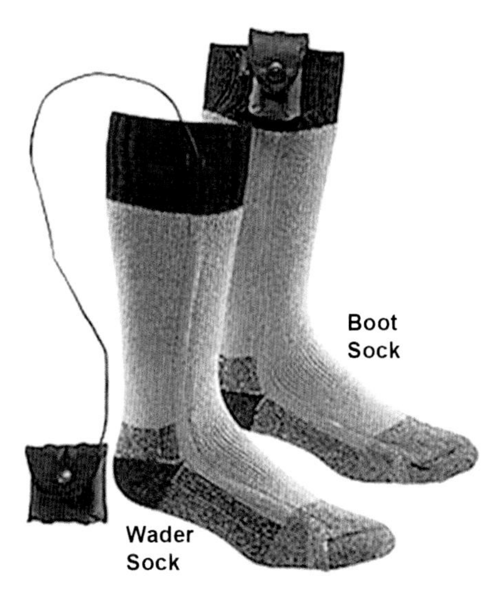

Fig. 7.2 Cabelas' battery powered Hunting Socks (Cabelas)

3. When the temperature has dropped, drink alcohol sparingly or not at all. Alcoholic drinks can cause the loss of body heat. Limit the amount of beer and wine. It's best that no alcohol is consumed.
4. Make sure to eat enough food to maintain body weight. A poor diet might result in having less fat under the skin. Body fat helps to insulate the body to stay warm.
5. Some illnesses may make it harder for the human body to stay warm. These include problems with the body's hormone system such as low thyroid hormone (hypothyroidism), health problems that keep blood from flowing normally (such as diabetes), and some skin problems where the body loses more heat than normal.
6. Use discretion during cold weather. Sometimes the skies are clear and it's extremely cold outside. Without a cloud layer to hold in the Earth's heat, a clear night allows the ground heat to radiate into the air and the environment becomes extremely cold. Sometimes it's not worth the risk of spending an extended time sitting at the telescope eyepiece.
7. Observe with a partner, friend, or companion just in case there is a health issue. If in the backyard, have someone check on you periodically. Don't go to remote locations alone.

Observing in the Fall and Winter

For many backyard astronomers, the winter is prime time for observing. The skies are dark and transparent because of low humidity. No insects are crawling up the legs or flying and buzzing around the head. The trees have lost their leaves and allow more sky to be seen. And rising from his summertime slumber, Orion rises above the horizon to welcome the astronomer to another wintry observing night.

But boy, is it COLD! There are challenges to amateur astronomy during the winter, all surrounding the fact that it's COLD! The two main challenges are keeping warm and keeping astronomy equipment functioning.

First and foremost, check the weather forecast and plan for dressing as if it is at least 20° colder than the weatherman predicts. Why? This is not a sporting event where a person is in constant motion, and consequently generating internal body heat. Astronomy is an activity that requires lots of sitting or standing at an eyepiece. Without proper precautions, all sorts of winter nasties can occur—shivering, frostbite, exposure, and worse. Dress in layers, wear a winter cap or hat, cover the ears, wear gloves, several layers of socks, boots, and long underwear....imagine Ralphie's little brother in the classic movie *A Christmas Story*. By dressing in layers, if it's too warm, just peel off a layer.

When it's cold outside, eat a good meal before going out into the field. Also, go to the bathroom before bundling up. Even if the observing site has a rest room or port-a-potty nearby, using the facility can be a challenge at night. Peeling down the layers to "eliminate" is a nuisance and will cause a loss of all the warmth built up.

Outdoor sports supply stores sell wonderful hand-warmer packs. These work. When hands get cold, these chemically activated packets will do wonders in warming up fingertips. Electric battery powered hand warmers are also available.

If the winter observing is done in the backyard at home, take breaks from the eyepiece to go inside and warmup. Just so long as it's dark inside the house to preserve night vision, there is no harm in warming up.

However, out in a dark country site, away from creature comforts, observe while the comfort level is good, but pack it in when the cold becomes a problem. Don't be a hero, this isn't a competition. Don't worry, the sky and the stars will be there on another night.

As far as the telescope equipment is concerned, there are a few preparations and precautions to take with cold temperatures of the winter. The cold weather rules for your telescopes are:

1. Let the telescope and eyepiece acclimate to the ambient temperature. Going from a warm home or car to a cool outdoor environment will require at least 30 min to adjust to the cooler air. If the temperature differential is greater, be prepared for a longer adjustment period. Open tube telescope designs, such as Newtonians and classic Cassegrain designs, will adjust to the ambient temperature readily. Closed tube designs such as refractors and catadioptric telescopes will take longer to adapt to the ambient air temperature.
2. Beware of dewing. As the temperature drops, optics can attract a layer of dew. There are two ways of combating dew: dew caps for the front of the telescope, and dew heater devices to gently maintain the temperature of the optics a few degrees above the dew point. Avoid observing objects directly overhead, since that position will accumulate dew on the optics. Once infected with dew, a brief exposure to the warm air of a hairdryer may help, but then the optics will have to acclimate to the ambient temperature all over again.

3. Rechargeable batteries should be fully charged. Fresh batteries should be installed if disposable batteries are used. Cold weather lessens the peak power of the batteries and shortens the length of use.
4. Beware of fogging, and condensation. It is easy to fog over eyepieces and finderscopes by inadvertently breathing on them. Don't breathe on your eyepieces or finderscopes.
5. Early autumn, the bugs and insects that bite and sting may still be a problem. As the weather gets colder, some remaining denizens may find a warmer home in eyepiece cases, telescope cases, and telescopes and mounts. Check all equipment before packing it in for a night.
6. If the user is mechanically inclined, special lubricants that work well at temperature extremes should be used on telescope mounts, focusers, and any other mechanical parts. Conventional oils and greases will thicken and become stiff in cold temperatures. Nothing is worse than being unable to focus the telescope because the focuser is frozen in place! If not mechanically inclined, seek out the nearest telescope dealer to help.

Heat Stress and Older Adults

The aging astronomer is particularly vulnerable to the problems of extreme heat, since all astronomy activities with a telescope are performed outdoors. Older bodies lose their ability to regulate temperature well. As with hypothermia, some medications may also interfere with the bodies' thermostat in adapting to high temperatures. Solar astronomers are especially affected during the heat of the summer. During some heat waves, even in the evening hours, the older astronomer must contend with warm temperatures and high humidity. A heat index in excess of 90–100° will stress the aging body. Heat exhaustion and the deadly heat stroke must be avoided by older people, especially for backyard astronomers.

Here are the different levels of heat stress to be watchful of and their treatments:

1. Heat rash, also called prickly heat, may occur in hot and humid environments where sweat cannot evaporate easily. When the rash covers a large area or if it becomes infected, it may become very uncomfortable. Signs and symptoms include a rash characterized by small pink or red bumps; irritation or prickly sensation; and itching. Heat rash may be prevented by resting in a cool place and allowing the skin to dry. To combat heat rash, keep skin clean and dry to prevent infection and wear loose cotton clothing. Cool baths and air conditioning are very helpful, and some over-the counter lotions may help ease pain and itching.
2. Sunburn is a familiar summertime malady. Solar astronomers are particularly vulnerable to sunburn. Redness, pain, and in severe cases swelling of skin, blisters, fever, and headaches can occur. There are ointments for mild cases if blisters appear and do not break. If breaking occurs, apply dry sterile dressing. Serious, extensive cases should be seen by physician.
3. Heat cramps (mild) are painful spasms usually in the muscles of the legs and abdomen, accompanied with heavy sweating. Use firm pressure on cramping

muscles, or gentle massage to relieve spasm. Give sips of water. If nausea occurs, discontinue use.

4. Heat cramps (prolonged) are painful muscle spasms that occur when someone drinks a lot of water, but does not replace salts lost from sweating. Tired muscles are usually the most likely to have the cramps. Signs and symptoms are cramping or spasms of muscles that may occur during or after the work. To treat heat cramps, drink fluids. A Gatorade or a similar sports drink can be consumed to replace the bodies' electrolytes. If the cramps are severe or not relieved by drinking a sports drink, seek medical attention.
5. Fainting usually happens to someone who is not used to working in the hot environment and stands a lot. Moving around, rather than standing still, will usually reduce the likelihood of fainting. Signs and symptoms include a brief loss of consciousness; sweaty skin, normal body temperature; and no signs of heat stroke or heat exhaustion. In a fainting situation, the victim should lie down in a cool place and seek medical attention if he or she has not recovered after a brief period of lying down.
6. Heat exhaustion can be recognized by heavy sweating, weakness, while the skin is cold, pale, and clammy. The victim's pulse becomes thready. Fainting and vomiting can occur. Get the victim out of sun, lay the victim down, and loosen their clothing. Apply cool, wet cloths. Fan or move the victim to an air-conditioned room. Give them sips of water. If nausea occurs, discontinue the water until it subsides. If vomiting occurs and continues, seek immediate medical attention.
7. Heat stroke (or Sunstroke) is the exact opposite of hypothermia, characterized by a high body temperature (103 °F or higher). Like hypothermia, heat stroke can be deadly. Watch the absence of sweating, with hot red or flushed dry skin, rapid pulse, difficulty breathing, strange behavior, hallucinations, confusion, agitation, disorientation, seizure, rapid and strong pulse leading to possible unconsciousness and coma. HEAT STROKE IS A SEVERE MEDICAL EMERGENCY. SUMMON EMERGENCY MEDICAL ASSISTANCE OR GET THE VICTIM TO A HOSPITAL IMMEDIATELY. DELAY CAN BE FATAL. Move the victim to a cooler environment. Reduce body temperature with cold bath or sponging. Use extreme caution. Remove clothing and use fans and air conditioners to aid the cooling process. If temperature rises again, repeat process. Do not give fluids. Persons on salt restrictive diets should consult a physician before increasing their salt intake.

Dressing for Success in Spring and Summer

As far as observing in the environmental conditions of spring and summer, here are a few things to remember:

1. Wear comfortable clothing. Spring clothing will tend to be heavier than summer clothing because of the slight nip in the springtime air. Summertime clothing

must be chosen to provide comfort in warm humid weather, but also provide some shielding from insects.

2. Use insect repellant. Some camping references recommend insect repellant with DEET. Some sporting goods or camping stores sell shirts and pants treated with an insect repellant or insecticide. Check with camping friends or even Consumer Reports magazine on the effectiveness of these treated clothes.
3. If possible, have an oscillating fan sweeping the observation area. Mosquitos are weak flyers and have difficulty finding their intended targets (i.e., you) when fighting against a wind (i.e., the oscillating fan).
4. Humidity may cause a decrease in the sky transparency. But the seeing conditions (the steadiness of the sky) will likely be excellent. Often, steady but slightly hazy conditions will yield good planetary observing.
5. The warmer weather and the resulting sweat that is generated by the human body can cause dehydration. Keep on hand plenty of liquid refreshments and stay hydrated. As tempting as it is, don't imbibe in alcoholic beverages. Drunk observing doesn't mix well with expensive eyepieces and telescopes.

Many of these environmentally caused health issues point to a common sense preventative and potentially lifesaving idea. Don't go observing without a partner or observing companion, especially to remote locations. The victim may not recognize the symptoms of hypothermia or heat stroke, but the companion will, and can take lifesaving action. Don't observe alone.

Chapter 8

Travel, Astronomy Tourism, and Getting Old

Airplane travel is nature's way of making you look like your passport photo.

—Al Gore.

Talk to any 55+ person who is planning for or is already retired about their future plans, and the majority will answer is to do a lot of traveling.

Leisure Travel and Astronomy

Travel is one of the most commonly listed interests of people, and it comes in all forms. Some people travel only when they need to visit relatives or friends, others travel on business, some travel for events (such as total solar eclipses, the Super Bowl, the Olympics, weddings, and honeymoons), and some travel as a sort of spiritual discipline, to discover new things about the world and its cultures and to learn more about themselves in the crucible that is travel. In the case of amateur astronomers, it's a chance to see a part of the sky that is unseen from their home (Fig. 8.1).

Yet for some, none of these things is that important. To these people, travel is about fun and excitement, about rest and relaxation, and about whatever they want it to be about.

Retirement is seen as a time to travel and take a vacation from everyday life. A leisure vacation is often characterized by staying in nice hotels or resorts, relaxing on beaches or in a room, or going on guided tours and experiencing local tourist

J.L. Chen, *Astronomy for Older Eyes*, The Patrick Moore Practical Astronomy Series, DOI 10.1007/978-3-319-52413-9_8

Fig. 8.1 Eta Carine, best seen in the Southern Hemisphere (Jon Talbot)

attractions. Most meals are eaten out when traveling for pleasure, and often more expensive modes of transportation, such as taxis, are used to get around. In some cases, leisure travel might be used to refer to any trip that lasts more than a week, regardless of the primary focus. Leisure travel is generally seen as the opposite of business travel.

This concept of leisure travel can be many things. Although many travelers spend more money than they would in their everyday life, others might choose to travel frugally. Backpacker travelers might stay in hostels, cook their own meals, and take public transit—as long as the main focus is on leisure, they are still leisure travelers. Many amateur astronomers wouldn't give a second thought to sleeping in a tent in the wilderness if it meant viewing the stars under a truly dark sky. Backpacking astronomers are often limited to carrying binoculars or small telescopes, but experiencing the dark skies of the wilderness makes it all worthwhile.

Similarly, a traveler might be seeking information for writing a book, making a new contact for information gathering, taking pictures to support the book, or work on a book proposal upon the return home (a slight autobiographical reference here!).

There are many benefits to travel. Stepping outside of a busy lifestyle can give people space to unwind and release their stress, returning to their everyday lives rejuvenated and refreshed. It also can give people the opportunity to experience parts of the world they have never seen before, enriching their understanding of other cultures, other peoples, and natural life in other parts of the planet. Traveling

Fig. 8.2 An aerial image of the NOAO's Cerro Tololo Inter-American Observatory in Chile. The largest dome is home to the Blanco 4-m telescope (NOAO/AURA/NSF)

can be an excellent way to meet new people and make new connections, and travel during the retirement years gives people the space and time they need to really forge those bonds. For amateur astronomers, travel is an opportunity to visit astronomy-specific places they have only read about in *Sky and Telescope* and *Astronomy* magazines (Fig. 8.2).

Even for those unwilling to embark on truly budget travel, there are many ways to limit costs. Some amateur astronomers who are retired have limited pensions or IRA savings to allow extravagant spending on travel. Fortunately, most popular leisure travel destinations offer a wide range in prices among hotels, restaurants, travel, and activities, depending on the season. Locations in the Caribbean or Mediterranean, for example, might be twice as expensive during winter, as millions of visitors from Europe and the United States try to escape cold weather.

A new development in the travel industry is the astronomy tourist. An astronomy tourist seeks to go to places either to observe the dark night skies through telescopes similar to those they own, or to visit major observatory sites.

The only caveat to astronomy travel is that taking trips becomes more difficult as one ages. Those in their 60s will find the traveling exhilarating and enjoyable. A noticeable slowdown occurs during the ages between 70 and 80 years when taking leisure trips for astronomy or just vacationing becomes difficult. After the age of 80, consult your doctor if travel can be safely accomplished. It is important to consider the availability of medical services at all destinations.

Some astronomy destinations are easily accessible. Some astronomy destinations are in rugged terrain, a challenge to get to, and are not necessarily for the fragile. Again, traveling at ages 50+ or 60+ is easier than traveling at 80+. Try to front load the more strenuous travel to the early stages of retirement, ages 55 through 65, when physical health is more likely to accommodate a challenging trip.

Top 10 Astronomy Destinations

The Smithsonian.com lists the following as the top 10 astronomy destinations to visit:

1. Mauna Kea Observatory, Hawaii, USA—13,796 ft above sea level and isolated in the Pacific Ocean, the observatories at Mauna Kea on Hawaii's Big Island offer some of the most pristine stargazing conditions in the world. Tourists can visit the summit, but officials suggest stopping at the Visitor's Center, located at 9200 ft, before continuing onward (both to check weather conditions and acclimate to the elevation). Many professional astronomers have to stop at this intermediate level in order to acclimate to the high altitude. Going to the observatories at the top of the mountain is stressful, and only for those of strong body. Check with your physician before the trip. Every night of the year, from 6:00 p.m. to 10:00 p.m., the Mauna Kea visitor center offers stargazing and sky tours to visitors, with telescopes available for amateur astronomers. It's completely free, and you don't need a reservation to participate. For a nominal fee, Star Gaze Hawaii sets up telescopes at hotels at sea level beach sites for nighttime observing (Figs. 8.3 and 8.4).

Fig. 8.3 Aerial view of Mauna Kea Observatory (Institute for Astronomy, University of Hawaii)

Fig. 8.4 Driving up to Mauna Kea Observatory (Michael Ficco)

2. Very Large Array, Socorro, New Mexico, USA—Fans of the movie *Contact* will recognize the Very Large Array, a massive radio telescope facility located 50 mi west of Socorro, New Mexico. The site is open for self-guided tours from 8:30 a.m. to sunset. On the first Saturday of each month, the facility holds free guided tours at 11:00 a.m., 1:00 p.m., and 3:00 p.m. No reservations are required for the guided tours, which run 30 min (Fig. 8.5).

   New Mexico is blessed with two astronomy resorts of note: StarHill Inn and New Mexico Skies. Both resorts offer an array of telescopes on their astronomy patios for rent, or guests can bring their own.
3. Royal Observatory, Greenwich, London, UK—Home of the prime meridian, the Royal Greenwich Observatory played a major role in the history of astronomy and navigation. Before the observatory was built, the grounds housed important buildings in English history dating all the way back to William I (the Tudors lived in Greenwich Castle, which was built on the same land as the Observatory). The Royal Observatory and Planetarium features a museum with a wide variety of exhibits (including a number about astronomical navigation techniques), as well as London's only planetarium (Fig. 8.6).
4. Cerro Paranal, Atacama Desert, Chile—Chile's Atacama Desert offers some of the most ideal stargazing conditions in the world: dry weather, cloudless skies, high altitude, and little to no light pollution. To experience the best this stargazing oasis has to offer, check out the Paranal Observatory, located on the

Fig. 8.5 The Very Large Array (NRL.navy.mil)

Fig. 8.6 The Royal Greenwich Observatory (RGO)

mountain of Cerro Paranal. Operated by the European Southern Observatory, Paranal is home to The Very Large Telescope, a grouping of four very large telescopes (over 320 in. in diameter). Guided tours of the observatory are offered to the public, without charge, every Saturday. Space is limited, so reservations are required. There are nearby hotels equipped to accommodate astronomy tourists, with private observatories for guests to reserve and use, or outdoor patios equipped with large amateur telescopes for guests to use (Fig. 8.2).

5. Kitt Peak National Observatory, Arizona, USA—The American Southwest offers some of the best stargazing conditions in the United States—and perhaps none are more choice than Kitt Peak, a national observatory southwest of Tucson, Arizona. Kitt Peak is home to the world's largest collection of optical telescopes, and offers guided tours daily at 10 a.m., 11:30 a.m., and 1:30 p.m. There are also nightly stargazing activities for those looking to peer at the cosmos through clear southwestern skies (Fig. 8.7).

   Kitt Peak Observatory has a 20-in. telescope which visitors and guests can reserve time on for observing or astro-imaging.

6. Griffith Observatory, Los Angeles, CA, USA—Los Angeles is known for its polluted and light polluted skies that might not offer the best stargazing conditions, but a visit to the Griffith Observatory is as much about the history of Los Angeles as it is about the stars. The Griffith Observatory was donated to the city of Los Angeles in 1896. It was also the location of an important scene in the movie *The Rocketeer*. The Griffith Observatory is open to visitors Tuesday through Sunday (Fig. 8.8).

Fig. 8.7 Kitt Peak National Observatory (National Optical Astronomy Observatory)

Fig. 8.8 Griffith Observatory (Griffith Observatory)

7. South African Astronomical Observatory, Sutherland, South Africa—At nearly 6000 ft above sea level sits the South African Astronomical Observatory, or SAAO, an observatory famous for its pristine sky conditions due to altitude and minimal air pollution. Located about 230 mi inland from the South Atlantic Ocean, the SAAO offers visitors a chance to tour their facilities and see telescopes that have been in operation since the 1970s. Visitors must call ahead and reserve space on a tour—the observatory offers two during the day, one fully guided (for 40 South African rands, or about $3.70) and one self-guided (for about $2.80). Night tours are also available on Monday, Wednesday, Friday and Saturday, during which visitors can look at the night sky through telescopes 14 in. and 16 in. in diameter (larger than what most amateurs would have the opportunity to use). Visitors cannot see any of the research telescopes during night tours, however.
8. Arcetri Astrophysical Observatory, Florence, Italy—If the hills of the Arcetri region of Florence were good enough for astronomy's ultimate bad boy (Galileo, maybe you've heard of him), then a visit to the Arcetri Astrophysical Observatory, located in the very same hills where Galileo spent the last years of his life, should be good enough for you too. Arcetri Observatory doesn't boast the massive telescopes of Kitt Peak or radio technology like the Very Large Array, but it offers a chance to step back in time to a historic period in astronomy. Daytime visits to the observatory are reserved for student groups, but nighttime visits are available for casual tourists. On Saturday evenings, the observatory holds open observatory sessions, where groups of up to five visitors are welcome to explore the observatory and grounds.

9. Teide National Park, Island of Tenerife, Canary Islands, Spain—In 2013, the Starlight Foundation, which works to preserve clear night skies, named Teide National Park, located on the island of Tenerife in the Canary Islands, both a "Starlight Reserve" and a "Starlight Tourist Destination," thanks to its pristine night skies and ideal stargazing conditions. Laws exist on the island to control light pollution and flight pathways in order to ensure perfect stargazing for visitors and astronomers alike. Tenerife is the site of a unique event called Starmus that combines astronomy, art, and culture. Tenerife is also home to one of the world's most advanced observatories, the Teide Observatory. Interested travelers can schedule visits (for a minimum of 15 people) by contacting the observatory.
10. Hayden Planetarium, New York City, USA—It's an unfortunate reality for star lovers that it can't be perfect weather all the time—sometimes it is cloudy and rainy, thwarting chances of a perfect starry night. For times like that, head to the Hayden Planetarium in New York City. Here, you can take in an IMAX or Space Show, or check out one of the Rose Center for Earth and Space exhibits: the Cullman Hall of the Universe, the Big Bang Theater (which features a show about the Big Bang), the Heilbrunn Cosmic Pathway, and the Scales of the Universe. Tickets to the Rose Center for Earth and Space and the planetarium must be bought through the American Museum of Natural History; general admission tickets start at $22 and offer access to the Natural History Museum as well as the space exhibits. Remember to say hello to Neil de Grasse Tyson while you are there! (Fig. 8.9).

U.S. State Department Travel Recommendations

As seniors, traveling outside of the United States is a new experience for many. Again, those in their 60s will find travel overseas an easier proposition than those in their 70s and 80s.

The U.S. State Department has alerts and warnings about certain areas of the world. Play heed.

The following is taken directly from the U.S. State Department website called Considerations for Older Travelers (any odd phases and poor grammar is not the author's fault!):

> "As an increasing number of older U.S. citizens are traveling abroad, the U.S. Department of State wants you to be prepared so you can enjoy your trip.
>
> 1. Make sure passports and/or passport cards are valid and have not expired.
> 2. Check to see if there is a Travel Warning or Travel Alert for your destination.
> 3. Check our Country Specific Information to determine if:
>   - You need a visa;
>   - You have enough blank pages in your passport for entry stamps;
>   - Your passport needs 6 months or more validity—some countries require this before they allow you to enter.

Fig. 8.9 Hayden Planetarium (Khayman)

Stay Connected

Enroll in our Smart Traveler Enrollment Program. Your information is stored securely and enables the Department of State, U.S. embassy, or U.S. consulate to contact you, your family, or your friends in an emergency according to your wishes.

Provide a copy of your itinerary, including contact information for where you will be staying to a friend or family.

Manage expectations—if you don't plan to stay in touch on your vacation, let your family know you will not be in regular contact.

Not all cell phones work abroad. If you want to have a cell phone with you as you travel you will need to check your cell phone coverage before you travel.

Health Information

Medical care in foreign countries varies and is often not up to U.S. Standards.

Medicare does not cover you overseas.

We highly recommend that you obtain medical and dental health care that will cover you overseas.

We also recommend medevac insurance.

For more tips related to health issues, visit our website. You may also find health information at the Travelers' Health page of the Centers for Disease Control and Prevention (CDC) website.

Pharmacies and Medications

Bring an ample supply of medication to cover you for your trip and if possible, for a few extra days in case there are delays. (Keep all medications in your carry-on, do not pack in check-on luggage—author).

Have information from your doctor regarding your condition and your medication.

To avoid questions or delays at customs or immigration, keep medications in their original, labeled containers.

Know the generic name for your medication as those generic names may be more recognizable at pharmacies in a foreign country.

Financial Information

Determine whether you should try to exchange currency before you travel abroad.

Understand the currency rates at your travel destination.

Know whether or not credit cards are readily accepted and if traveler's checks can be cashed and plan accordingly.

Make sure your credit card company knows you will be traveling abroad so they do not freeze your accounts.

Read the Crime section of the Country Specific Information for the countries you will visit to review the ATM scams and other financial scams that may be targeting foreign visitors. If ATM service is not widely available or is not secure, bring travelers checks and one or two major credit cards instead of planning to use cash. Many banks in most countries will issue cash advances from major credit cards.

Accessibility and Accommodations

For more information, check our section on Traveling with Disabilities.

Beware of Scams

Scammers intend to get money from their victims by making the victims believe they will gain something of great personal value (financial gain, a romantic relationship, helping someone in trouble, the safe return of a friend, etc.).

Scammers operate primarily via the Internet, email, and phone. For more information, please review our information on International Financial Scams.

Prepare for Emergencies

Leave emergency contact information and a copy of your passport biographic data page with family and trusted friends.

Carry emergency contact information for your family in the United States with you when you travel (be sure to also pencil it in the emergency contact information section of your passport).

Know the contact information for the nearest U.S. embassy or consulate, available on the Country Specific Information page for each country and on each embassy or consulate's website, and provide that information to your family and friends.

If there is an emergency situation where you are staying, such as civil unrest, disrupted transportation, or a natural disaster, prevent undue worry or concern by contacting your family and friends as soon as possible.

A secure way to maintain your emergency contact information is to enroll with our Smart Traveler Enrollment Program.

Take careful note of the cancellation policies for your travel and consider purchasing travel and luggage insurance. Many credit card, travel, and tourism companies offer protection packages for an additional fee.”

Chapter 9

Common Sense, Light Pollution, and Astronomy

"...as light pollution spreads,
we are slowly losing one of the oldest and most universal links
to all of humanity."

—Peter Lipscomb, Santa Fe Astronomer

Buying a Telescope

A useful analogy in buying a telescope is looking at a parking lot of a local grocery store. There are a variety of cars and trucks parked there. Why? Because different people purchase vehicles for different reasons. Soccer moms need minivans to haul their kids to soccer fields. Handymen need pickup trucks to haul plywood and plumbing tools. The thrill-seeker will own a high-performance sports car. And a business man will drive a prestige high-priced car to show off wealth and fame. Many amateur astronomers own multiple telescopes to fulfill specific observing needs (Fig. 9.1).

The same process of selection also applies to telescopes. In this case, there are telescopes that are best used for deep sky objects such as nebulas, galaxies, and star clusters. There are telescopes that excel in astrophotography. There are telescopes that excel in observing the Moon and the planets of our solar system (Fig. 9.2).

First-time buyers are faced with a myriad array of telescope choices, and more-often-than-not purchase the wrong telescope for their use. The wrong telescope purchase will end up in the closet gathering dust, or worst yet, in a garage sale.

J.L. Chen, *Astronomy for Older Eyes*, The Patrick Moore Practical Astronomy Series, DOI 10.1007/978-3-319-52413-9_9

Fig. 9.1 Celestron SCTs at NEAF 2016 (James Chen)

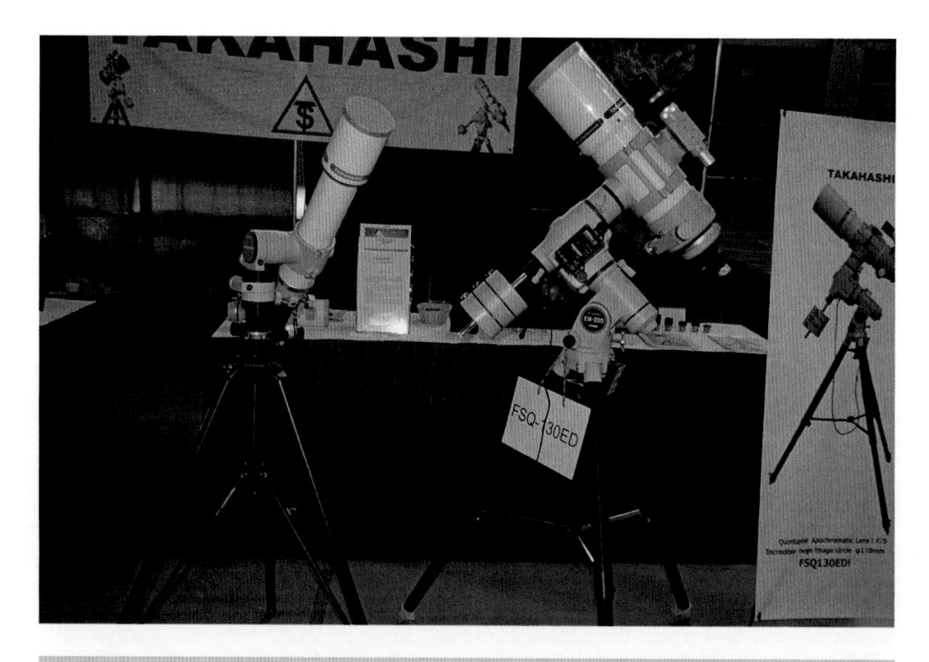

Fig. 9.2 Takahashi telescopes at NEAF 2016 (James Chen)

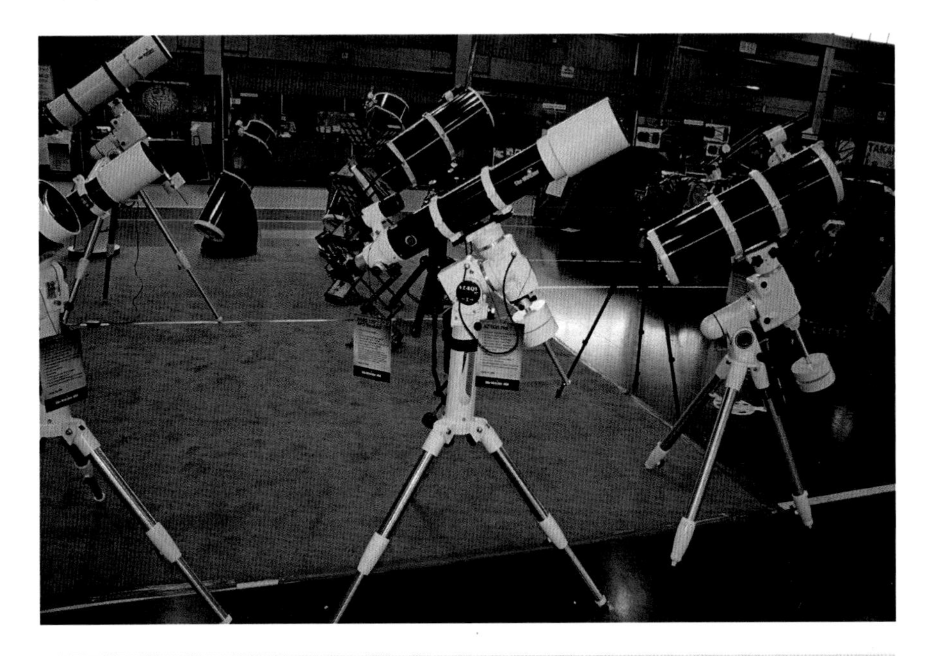

Fig. 9.3 Sky Watcher booth at NEAF 2016 (James Chen)

So here are a few basic all-encompassing guidelines in selecting telescopes for astronomy use (Figs. 9.3 and 9.4).

1. *Buy your second telescope first.* The common advice for years from all amateur astronomers is don't buy a department store telescope. In today's world, that advice extends to warehouse stores and sporting goods stores. Most so-called beginners' telescopes are plagued with poor optics, shaky telescope mounts, and in some cases poor electronics. Many of these telescopes are aimed at well-intentioned consumers that haven't taken the time to study the telescope market, and just want a big box under the Christmas tree or at the birthday party. Grandparents especially fall into this trap. By using the term second telescope, most telescope owners who survive the trials of these beginners' telescopes and still want to pursue the hobby naturally learn to buy a quality telescope the second time around. Save money now by being educated and buy the right equipment first.
2. *A smaller telescope will get used more than a larger telescope.* There is a strange ailment that afflicts every backyard astronomer known as aperture fever. In this bigger-is-better society, the desire for a larger telescope that shows more detail and gathers more light is sometimes overwhelming. But there is a point where a telescope becomes so large and cumbersome to use that the usage of said telescope becomes less and less. A smaller and more portable telescope with easy setup gets used more.

Fig. 9.4 Vixen Twin-telescope binoculars at NEAF 2016 (James Chen)

3. *The telescope mount is as important as the telescope optics.* A good, solid, and stable telescope mount encourages observers to use their telescope. Nothing is more frustrating than trying to focus a telescope on a weak and poorly designed mount that shakes and vibrates with a slight touch or a slight breeze.
4. *The right eyepieces for the right job.* As with telescope designs, certain eyepiece designs are suited for wide-angle extended celestial objects such as nebulas and open star clusters, while others are intended to high contrast detailed assignments. With the cost of eyepieces ranging from $30 to over $1000 each, a meaningful and careful selection is appropriate.
5. *Buy the right telescope that suits your personal skills.* Some telescopes are well suited for the technically inclined. Some telescopes are simple to use. The potential first-time telescope owner needs to understand their own personal skills and acknowledge their abilities before making a telescope selection. Namely, if you can't change a car tire, or your digital oven clock is always flashing 12 o'clock, certain hi-tech astro equipment should be avoided. And as with the size of the telescope, the easier the telescope is to use, the more likely it will be used.
6. *Consider a neutral density or polarizing filter for the telescope eyepiece.* The Moon, especially when it's full, can get uncomfortably bright. Not dangerously bright, like the Sun, just uncomfortable. An appropriate filter will tone down the glare to a comfortable level.

7. **CAUTION: DO NOT OBSERVE THE SUN WITHOUT PROPER SOLAR FILTERING EQUIPMENT. This is a repeat from Chapter 2, but is important:**

   Never look directly at the Sun with the naked eye or with a telescope, unless the proper solar filter is being used. Permanent and irreversible eye damage will result without proper protection. Solar observing does demand a different suite of equipment, including white light solar filters, H-alpha filters, and dedicated H-alpha telescopes.

   Never use the telescope to project an image of the Sun onto any surface. Internal heat buildup can damage the telescope and any accessories attached to it.

   Never use an eyepiece solar filter with a telescope. Internal heat buildup inside the telescope can cause these devices to crack or break, allowing unfiltered sunlight to pass through to the eye and cause irreparable damage and blindness.

   Never leave the telescope unattended when viewing the Sun. People and children unfamiliar with the dangers of viewing the unfiltered Sun may do something stupid if left alone with the telescope. Never underestimate the dumbness and stupidity of the general public.
8. *Buy quality.* The old adage "You get what you pay for" applies here. Telescope images are clear and sharp. Mounts work smoothly and are free of spurious vibrations. Focusers have a buttery smoothness that allows for fine tuning of the focus. High-quality telescopes allow the observer to enjoy astronomy without problems getting in the way. In fact, there are numerous examples of quality apochromatic refractors that have appreciated in value, and sell on the used market for more than the original purchase price.
9. *Support your local telescope store.* Believe it or not, the astronomy industry is not a big money, high profit business. With the exception of two dominant major companies, many telescope businesses, either manufacturers or stores, are Mom and Pop operations run by people who love science and astronomy. They have expertise in amateur astronomy, provide quality products, provide personalized service, and are able to perform many repairs in their own shops. The smaller telescope shops struggle to compete with high volume Internet or mail-order firms who offer little or no service and rely on manufacturers to repair faulty equipment. Consumers need to understand the retail business. There are three criteria for competition: Quality, Service, and Price. The consumer can only get two of the three. A lower price means the consumer sacrifices either service or quality. It is astounding to note that profit margins of major name brand telescopes are miniscule. For example, a well-known large Schmidt–Cassegrain computer-controlled telescope costing over $3000 will net a profit to a store of $100. Smaller stores rely on accessory sales, service work, and loyal customers to stay in business. Remember, at your local telescope store, there are real people (not a disembodied voice on the phone) who know astronomy, sell and support quality products, support local astronomy clubs, and can fix any problems with telescope equipment (often on the spot).

10. *Avoid waiting lists for new telescope equipment.* There is an anecdotal story that bears repeating of the 70-year-old man who was shopping for a cutting edge quality apochromatic telescope in the Washington, D.C. area about a decade ago. This gentleman could have obtained a 102 mm Japanese-made high-quality apochromatic refractor and received delivery within 2 weeks of placing his order. Instead, he chose to place an order for a similar size telescope with a well-known American refractor manufacturer with a waiting list of 3 years. Under normal circumstances, this would not be an unreasonable decision. But at age 70, with the expected delivery of the ordered telescope when he is at age 73, this decision was questionable. Price was not an issue. The performance comparison between the two telescopes revealed them to be arguably equivalent. The question arises: Why wait 3 years, with declining eyesight, weakening health, declining physical strength, and the inevitable is closer at hand? We're not getting any younger!

Observing and the Importance of Dark Skies

As one grows older, the comforts of home are very alluring. An amateur astronomer must find a balance between living in a comfortable house, having good neighbors and friends, a friendly community, and the need to have dark skies and live away from light pollution. Often, one's home is situated where one portion of the sky is darker and more observable than the rest of the sky. Sometimes, the dark sky solution is to load the car with one's telescope equipment and drive to a star party or dark location in the country. Or, as this author did, buy a house in the country and away from city lights and have dark sky all the time!

As seen in previous chapters, the aging human eyes become less sensitive to light under low level light conditions. Light pollution's stray light from car headlights, streetlights, parking lot lights, business signs, and house outdoor lights add to the challenge for the aging astronomer's ability to view the stars.

A 2016 study documented light pollution found across the globe that more than 80% of the world's population lives under light polluted skies. Conditions in the United States and Europe are even worse, with 99% of their citizens experiencing skyglow at night.

Light pollution is the result of outdoor lighting that is not properly shielded, allowing light to be scattered into the night and into the human eye, and disrupting the darkness of the night sky. Light pollution changes the night sky from being dark and full of stars to a grey night sky with few stars visible. Poor light that shines into the eyes is called glare and the light shining into the night sky above the horizon causes skyglow. Poor lighting can also cause light trespass when it is directed into areas that it is not wanted, e.g., a neighbor's yard and windows.

Many astronomers choose to travel to dark sites in order to see the Milky Way and deep sky objects. Some (as this author) have chosen to move to a dark country home far away from the disruption of skyglow caused by street lights, business signs, and parking lot lighting (Fig. 9.5).

Fig. 9.5 Comparison of the rural night sky and an urban night sky (Jeremy Stanley)

Some amateur astronomers have chosen to tackle light pollution head-on by approaching politicians, business owners, and communities to changing their nighttime lighting practices, often with the support of the IDA. The International Dark-Sky Association (IDA) is a United States-based nonprofit organization with the mission to preserve and protect the nighttime environment and the heritage of dark skies through quality outdoor lighting. IDA was the first organization in the dark sky movement and is currently the largest (Fig. 9.6).

Fig. 9.6 A comparison of the Orion constellation under darks and light polluted sky (Jeremy Stanley)

The impact of light pollution affects astronomers in a major way. In addition to being unable to view the night sky, light pollution is also detrimental in other ways, as described on the IDA website:

1. Energy waste—Lighting is responsible for one-fourth of all electricity consumption worldwide, and case studies have shown that several forms of over-illumination constitute energy wastage, including non-beneficial upward direction of nighttime lighting. The question has been raised by many: Why are we spending so much energy and money lighting up the undersides of airplanes and the bottoms of night flying birds and bats?
2. Effects on animal and human health and psychology—Medical research on the effects of excessive light on the human body suggests that a variety of adverse health effects may be caused by light pollution or excessive light exposure, and some lighting design textbooks use human health as an explicit criterion for proper interior lighting. Health effects of over-illumination or improper spectral composition of light may include increased headache incidence, worker fatigue, medically defined stress, decrease in sexual function, and an increase in anxiety levels.
3. Disruption of Ecosystems—Ecological light pollution affects organisms and ecosystems. While light at night can be beneficial, neutral, or damaging for

individual species, its presence invariably disturbs ecosystems. Light pollution poses a serious threat in particular to nocturnal wildlife, having negative impacts on plant and animal physiology. It can confuse animal navigation, alter competitive interactions, change predator–prey relations, and cause physiological harm. The rhythm of life is orchestrated by the natural diurnal patterns of light and dark, so the presence of light pollution disrupts these patterns and impacts the ecological dynamics.

4. IDA focus on sea turtles—Sea turtle hatchlings emerging from nests on beaches are another casualty of light pollution. It is a common misconception that hatchling sea turtles are attracted to the moon. Rather, they find the ocean by moving away from the dark silhouette of dunes and their vegetation, a behavior with which artificial lights interfere.
5. Impact of light pollution on other wildlife—Juvenile seabirds may also be disoriented by lights as they leave their nests and fly out to sea. A Scientific American reported in the July 2016 issue that city-dwelling moths have evolved to avoid lamps and city lighting in an adaptation to light polluted areas. Amphibians and reptiles are also affected by light pollution. Introduced light sources during normally dark periods can disrupt levels of melatonin production. Melatonin is a hormone that regulates photoperiodic physiology and behavior. Some species of frogs and salamanders utilize a light-dependent "compass" to orient their migratory behavior to breeding sites. Introduced light can also cause developmental irregularities, such as retinal damage, reduced sperm production, and genetic mutation.

The IDA is working to reduce light pollution, such as reducing skyglow, reducing glare, reducing light trespass, and reducing clutter. The method for best reducing light pollution, therefore, depends on exactly what the problem is in any given instance. Solutions include:

1. Utilizing light sources of minimum intensity necessary to accomplish the light's purpose.
2. Turning lights off using a timer or occupancy sensor or manually when not needed.
3. Improving lighting fixtures, so that they direct their light more accurately towards where it is needed, and with fewer side effects.
4. Adjusting the *type* of lights used, so that the light waves emitted are those that are less likely to cause severe light pollution problems. Mercury, metal halide and above all first generation of blue-light LED road luminaires are much more pollutant than sodium lamps. The Earth's atmosphere scatters and transmits blue light better than yellow or red light. It is a common experience to observe "glare" and "fog" around and below LED road luminaires as soon as air humidity increases, while orange sodium lamp luminaires are less prone to show this phenomenon.
5. Evaluating existing lighting plans, and redesigning some or all of the plans depending on whether existing light is actually needed (Fig. 9.7).

Fig. 9.7 International Dark-Sky Association Acceptable and Unacceptable Lighting Fixtures (IDA)

Often, the detrimental effects of light pollution go unrecognized, as the environmental impacts appear to the general public as "in the noise" and the public's perception of the information is minimal. But like the case of climate change, the accumulation of the scientific evidence is overwhelming.

Modern society requires outdoor lighting for a variety of needs, including safety and commerce. IDA recognizes this but advocates that any required lighting be used wisely. To minimize the harmful effects of light pollution, lighting should:

1. Only be on when needed.
2. Only light the area that needs it.
3. Be no brighter than necessary.
4. Minimize blue light emissions.
5. Be fully shielded (pointing downward).

The illustration below provides an easy visual guide to understand the differences between unacceptable, unshielded light fixtures and those fully shielded fixtures that minimize skyglow, glare, and light trespass.

Most people are familiar with incandescent or compact fluorescent bulbs for indoor lighting, but outdoor lighting usually makes use of different, more industrial, sources of light. Common light sources include low-pressure sodium ("LPS"), high-pressure sodium ("HPS"), metal halide, and light emitting diodes ("LEDs").

LPS is very energy efficient, but emits only a narrow spectrum of pumpkin-colored light that some find to be undesirable. Yet, LPS is an excellent choice for lighting near astronomical observatories and in some environmentally sensitive areas.

HPS is commonly used for street lighting in many cities. Although it still emits an orange-colored light, its coloring is more "true to life" than that of LPS.

In areas where it's necessary to use white light, two common choices are metal halide and LEDs. One of the advantages of LED lighting is that it can be dimmed. Thus, instead of always lighting an empty street or parking lot at full brightness, LEDs can be turned down, or even off, when they aren't needed and then brought back to full brightness as necessary. This feature both saves on energy and reduces light pollution during the night.

Because of their reported long life and energy efficiency, LEDs are rapidly coming into widespread use, replacing the existing lighting in many cities. However, there are important issues to consider when making such a conversion.

As in the illustration above, it is crucial to have fully shielded lighting, but we now know that the color of light is also very important. Both LED and metal halide fixtures contain large amounts of blue light in their spectrum. Because blue light brightens the night sky more than any other color of light, it's important to minimize the amount emitted. Exposure to blue light at night has also been shown to harm human health and endanger wildlife. The IDA recommends using lighting that has a color temperature of no more than 3000 K.

Lighting with lower color temperatures has less blue in its spectrum and is referred to as being "warm." Higher color temperature sources of light are rich in blue light. IDA recommends that only warm light sources be used for outdoor lighting. This includes LPS, HPS, and low-color-temperature LEDs. In some areas, the white light of even a low-color-temperature LED can be a threat to the local nighttime environment. In those cases, LPS or narrow-spectrum LEDs are preferred choices.

The IDA has instituted a Fixture Seal of Approval program to encourage cities, municipalities, communities, and individuals to purchase and install the proper outdoor lighting fixtures that are night sky friendly.

There are many solutions that an older amateur astronomer can take in battling light pollution:

1. Become an advocate in the community seeking light pollution changes and solutions.
2. Find a convenient dark sky area to transport telescope equipment and observe.
3. Join an astronomy club to team with like-minded people to fight the ravages of light pollution.
4. Be an educator on light pollution.
5. Write to city administrators, state legislators, congressmen, and senators about light pollution.
6. Write articles in community newsletters and local newspapers on the subject of light pollution.

Chapter 10

Wheelchair Astronomy

Sometimes I wonder if I'm as famous for my wheelchair and disabilities as I am for my discoveries.

—Stephen Hawking

Unfortunately, a few aging astronomers will find themselves in a wheelchair. Bummer! But this shouldn't stop anyone's participation in observing the stars. It just takes a little adaptation, cleverness, and a few trade-offs. Keep in mind, the seated position is still preferable to standing when viewing through a telescope. The only difference is this chair has wheels, and for whatever reason, the observer can't leave it.

As any disabled reader will know, accepting physical limitations is one of the hardest parts of dealing with physical disability. Not being able to do things as was done before requires lifestyle adjustments.

Compromises have to be made. There is no point in purchasing that huge GoTo 12-in. SCT, a large 24-in. Dobsonian that needs a 10-ft ladder to access the eyepiece, or a 152 mm refractor with 40 pounds of counter weights for the German equatorial mount. A wheelchair-bound astronomer will never be able to move, set up, and align these monsters of optics and machinery.

To help understand the accommodations needed for wheelchair astronomy, the following three scenarios are presented here. Many of the recommendations and considerations overlap the scenarios, but common to all three is the process of thinking through the restrictions to accommodate a wheelchair-bound astronomer.

J.L. Chen, *Astronomy for Older Eyes*, The Patrick Moore Practical Astronomy Series, DOI 10.1007/978-3-319-52413-9_10

Scenario 1—Wheelchair Astronomer with a Caregiver

In most cases, the disabled person will have a wife, husband, significant other, or nurse that provides aid to everyday life for the wheelchair bound, i.e., the caregiver.

Of the three scenarios to be discussed, the extra hands and legs of the caregiver makes life for the wheelchair astronomer conceivably simpler and easier. The ability to observe through a telescope for the wheelchair astronomer may not change at all, with the exception that the caregiver now performs the moving and setting up of the telescope equipment. This must be discussed, negotiated, and agreed upon between the caregiver and the wheelchair astronomer.

Be prepared to downsize, especially if the caregiver does not share the same level of interest for the night sky. Hauling around and setting up a 14-in. SCT versus picking up a 4-in. Maksutov requires two distinct effort levels. One needs the strength of a weight lifter and the other is as easy a moving a dinner chair.

When downsizing, select telescopes and their mounts that are of appropriate size and can be easily moved. So that means the mount, OTA, diagonal, eyepieces, and accessories have to be thought of in terms of portability and weight. When faced with a choice, err on the side of caution and go as light as possible. A grab-and-go telescope is an appropriate choice.

For example, if the previous telescope was a 12.5-in. Newtonian, 12-in. SCT or larger telescope, consider downsizing to a 90 mm refractor or a 6-in. SCT. A smaller sized telescope is easier to move and set up. The smaller sized telescope will probably be more acceptable to the caregiver. Additionally, the eyepiece location on the telescope will be more accessible.

Scenario 2—Wheelchair Astronomer Alone

There are many wheelchair-bound astronomers who are able to live independently. Aperture compromise is particularly necessary in choosing a telescope in this scenario.

Conceptually, the most adaptable telescope for the wheelchair-bound is the traditional grab-and-go sized telescope. Small in size and high in performance, many aging astronomers will most likely own a grab-and-go telescope already. The following options are recommended:

1. Binoculars—As noted earlier in this book, binoculars are often the forgotten astronomy tool. A decent pair of 10 × 50 or 12 × 60 binoculars gives excellent wide-field views of the night sky. Prolonged observing with binoculars may necessitate some form of tripod and support, even though binoculars are lightweight, easy to transport, and easily used from a wheelchair. Coupled with either a specialist mirror type mount or a parallelogram binocular mount, a person in a wheelchair can comfortably use 10 × 50 or larger sizes, only being limited by

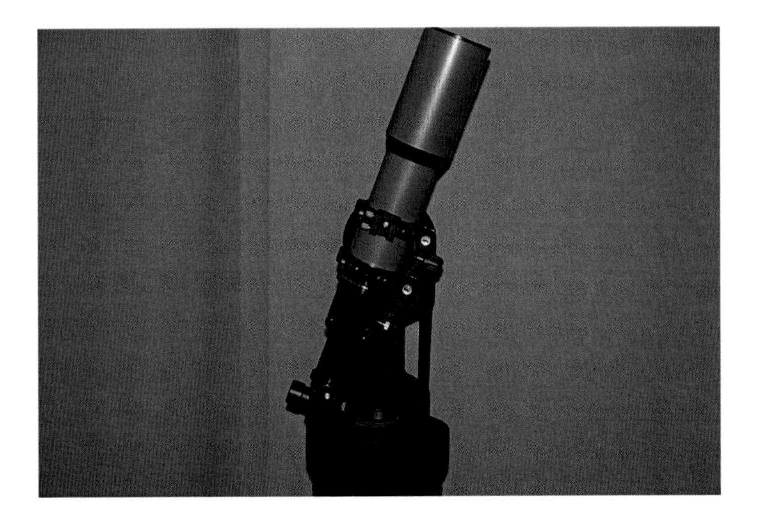

Fig. 10.1 Short-tube Stellarvue SV80 on a Celestron GoTo mount (James Chen)

what one can safely carry. A custom mounting for wheelchair users can be fashioned to accommodate a parallelogram type mount onto the back of a wheelchair.

2. Short focal length refractors up to 90 mm diameter—There are a number of 66 mm, 70 mm, 80 mm, and 90 mm short focus refractors on the market. These grab-and-go telescopes are the ideal size for wheelchair applications. The quality of the small apochromatic refractors makes them adaptable for use as world-class imaging telescopes when coupled to either a CCD camera or a digital SLR camera (Fig. 10.1).
3. Small Maksutov—Small Maksutov telescopes up to 90–102 mm can be managed on a lightweight alt-az or driven fork mount.
4. Small SCT—It's very rare to find an SCT of 4″ or under these days as most have been replaced with the Maksutov design. There are small 5″–6″ SCTs available on the market with full GoTo capabilities that should be considered for wheelchair applications (Fig. 10.2).
5. Tabletop Telescopes—Several online sites have discussed the use of tabletop telescopes, either Maksutov or SCT designs (Fig. 10.3). At first glance, this seems an excellent option, but the type of table the telescope is set upon becomes critical. There are stability issues of the table to consider. Flimsy card tables need not apply here. Wooden picnic tables may be used for telescope application only if wheelchair accessibility can be accomplished. Access to all positions around the table to accommodate the various eyepiece positions that will be encountered as the telescope slews from one object to another must be considered. A solid round table appears ideal, but the positioning of the tables leg support must not interfere with the wheelchair (Fig. 10.3).

Fig. 10.2 6″ Celestron evolution 6 on a GoTo mount (Celestron)

The choice of telescope mounts is restricted, with German equatorials being too heavy to manage from a wheelchair. There are a handful of desktop equatorials that may be considered.

1. German Equatorial Mounts—German equatorials are best left to a permanent installation. Just the sheer weight of the counterweights for German mounts deters this mount design from consideration.
2. Altitude-Azimuth (Alt-Az)—For purely visual use, a good alt-az mount is highly recommended, either the ones with slow motions or the balanced push-type. There are many choices out there worth investigating at every price point (Fig. 10.4).
3. Photographic Tripods—Some photographic tripods may also be considered. A good solid metal one will carry most small scopes and, while not the best solution, is more than adequate to start with.
4. Major telescope manufacturers have several telescope models available with GoTo fork mounts, an example is seen in Fig.10.2.

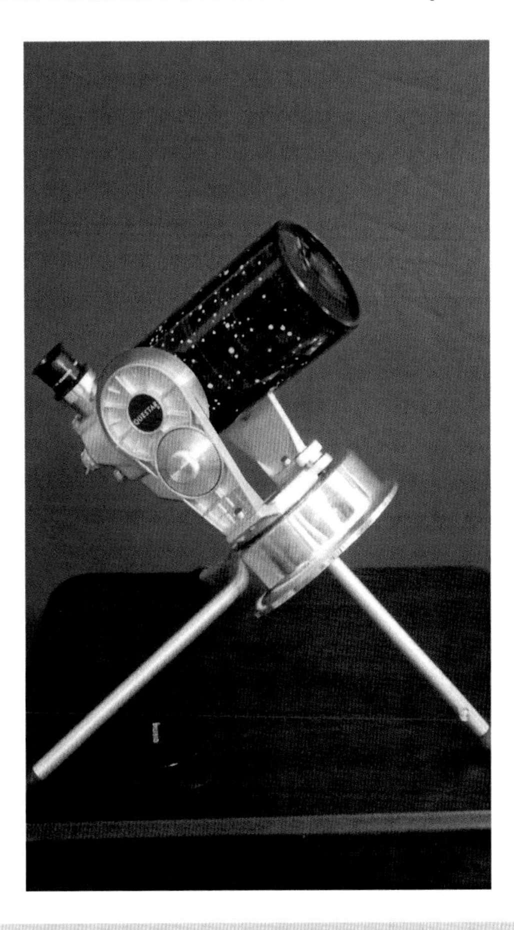

Fig. 10.3 Classic Questar 3.5-in. tabletop telescope (Hands-on-Optics photo archives)

Scenario 3—Wheelchair Astronomer with Additional Disabilities: Open the Window

There are a large number of disabled people who cannot get outside for a variety of reasons. Most assume that if they can't get out, they can't take up astronomy. There are challenges, but it is possible to view the stars from indoors.

Looking out of windows is not the norm. Most astronomy books advise against this because of the thermal difference that can be present from the internal home temperature and the outside ambient air which can cause air turbulence around the window. But if done correctly, the view from a telescope pointed through a window can fulfill the need to see the stars. Opening a window up, poking the telescope through it and observing is not commonly recommended. So long as the window is not obstructed by trees or other buildings, this option can be considered.

Fig. 10.4 Author's Stellarvue SV80BV on a Vixen Porta II alt-az mount (James Chen)

There are of course a few simple rules in order to gain the best from this method:

1. Turn ALL the lights off; turn off the TV and computer monitor as well. This will help with getting the eyes dark adapted.
2. If you have a door or window that opens in the direction you want to look, open it, and look through it with the naked eye. Remove any window screens. Make sure there are no trees obstructing the view. If it's winter, the trees will have lost their leaves, which helps. It's possible to observe the night sky between the tree branches!
3. Turn off the heating or air conditioning to the room and allow the room time to come to ambient temperature.
4. Open the window and point the telescope through the window. Refractors or SCT telescopes are suggested as good candidates for this application.
5. The freedom of all-sky viewing will be hampered, but the sky view through the window should allow viewing of planets, star clusters, and star fields. The dimmer objects may be challenging because the house is radiating its thermal energy into the night air and causing turbulence.
6. Don't use short focal length eyepieces; this is not an occasion to push the magnification. The high magnification may be challenging because the house is radiating its thermal energy into the night air. There will be a shimmering effect on celestial objects at high magnifications.
7. Make sure you are warm enough and comfortable.

Though the disabled astronomer has to take account of certain limitations such as size and weight of the instrument as well as the areas he or she has access too, there are still plenty of options there that allow you to see our wondrous universe.

With a little thought, preparation, and often a little ingenuity, most of the obstacles can be overcome.

Observatory

Depending on the budget, a permanent observatory is a consideration (Fig. 10.5).

This could be a new-build structure, a shed converted into an observatory, or even just a mount outside with a good quality all weather cover to protect it from the adverse weather. Or the tripod that comes with the mount can be left outside. Precautions must be made to protect the mount mechanisms from the weather, and particularly moisture. Whatever you decide to get make sure it's secure, weather proof, and secured from theft.

If the location is secure enough and there is a comfort level about leaving an expensive telescope permanently mounted then, as far as optics go, the sky's the limit.

An 8″–12″ SCT on a GoTo Equatorial mount will really open up the skies. Add an electronic eyepiece or CCD camera, remote control for the mount, and cabling to provide images to be transmitted indoors, observing can be performed much like

Fig. 10.5 The late Dr. Charles Leeper's DIY observatory (James Chen)

the professional astronomers do, from a computer monitor. This is a different experience than actual eyeball observing, but remember the operative word is compromise.

The main points to remember are:

1. Identify and accept the physical limitations, work within them not against them.
2. Whatever the temptations stick with what can be easily managed without help.
3. Buy the most solid mount that can be readily managed.
4. Don't rule out camera tripods if that's all that can be carried. (It is far better to have a lighter mount that can be managed than a larger one that can't.)
5. If going for a permanent mount, ensure it's secure, weather proof and alarmed if possible.
6. There will be maintenance issues with owning an observatory. As with owning a home, the observatory will need upkeep, as in painting, caulking to prevent water leaks, lubrication of observatory bearings, cleaning to remove dust, dirt, spiders, ants, etc. Arrange for someone to aid in the upkeep of the observatory.

Observing Site

A little common sense goes a long way when it comes to choosing observing sites, either at home or at a remote site. The able bodied will only have to worry about whether or not the ground will support their equipment. The wheelchair-bound also have to think about the nature of the surface and its risks:

1. A nice dark site situated in a green field a few miles from home is a lovely prospect, but no good at all if a wheelchair cannot traverse it. Wheelchairs are not SUVs or all-terrain vehicles.
2. Look into parking access and choose areas with good solid ground and as few obstacles as possible. A parking lot, terrain wise, is ideal, except most parking lots are well-lit light pollution disasters.
3. Wheelchair accessibility is the goal in choosing where the telescope can be set up safely. For a home site, consider lighting the path to the site with lights that can be turned off. Safety is a high priority.
4. Try to keep things at ground level and on a well-drained part of the site. No steps, bumps, or divots are allowed. No muddy terrain either.
5. If there is a favorite area of the backyard, then it is worth considering getting a concrete slab laid and leveled if there isn't one there.
6. A simple ramp and hand rail leading onto it is all that's needed to make it easy to get on and off, and helps even those who don't use a wheelchair.

If going to a star party, then make sure the organizer is informed of the disability and reserves a suitable spot. Make sure the sponsor club members are available to help, if needed.

A little pre-planning and common sense is all that's needed to ensure an enjoyable night at the eyepiece.

Chapter 11

The Afterlife of Telescope Equipment and Astronomy Books

Customer: 'E's not pinin'! 'E's passed on! This parrot is no more! He has ceased to be! 'E's expired and gone to meet 'is maker! 'E's a stiff! Bereft of life, 'e rests in peace! If you hadn't nailed 'im to the perch 'e'd be pushing up the daisies! 'Is metabolic processes are now 'istory! 'E's off the twig! 'E's kicked the bucket, 'e's shuffled off 'is mortal coil, run down the curtain and joined the bleedin' choir invisible!!

THIS IS AN EX-PARROT!!

—Monty Python in the Dead Parrot Skit

Like Monty Python's dead parrot, seen in Fig. 11.1, our lives will take their course. But the lifetime of telescope equipment can last for several generations. 's telescope still exists as a museum piece. Alvin Clark's 1873 26″ refractor at the U.S. Naval Observatory still remains in use for astrometric measurement. Antique Mogey, Brashear, Broadhurst, Clarkson, and Fuller telescopes are widely sought by collectors of antique scientific instruments. Zeiss telescopes, along with Golden Age telescopes by Cave Optical, Optical Craftsman, and Unitron are highly prized instruments that are still continually used for their intended purpose—seeking and gathering faint starlight.

There is value to these instruments that often our family members don't appreciate during our lives, and as heirs to our property, sometimes won't understand the telescope's true value.

Discussions with our family in matters of the estate and inheritance are difficult to conduct. Lawyers, accountants, and financial planners often recommend that estate planning begin as early as possible. As uncomfortable as they can be, family talks must be conducted not only in the disposition of your telescope equipment,

J.L. Chen, *Astronomy for Older Eyes*, The Patrick Moore Practical Astronomy Series, DOI 10.1007/978-3-319-52413-9_11

Fig. 11.1 Deceased "Norwegian Blue" (Twitter)

but also for the disposition of the rest of your estate. Proceed thoughtfully and with the guidance of a lawyer, estate planner, and a certified public accountant. If done well, the inheritance tax implications can be minimized or completely avoided. Seek a professional on these matters. Your Uncle "Fred" or Aunt "Mary" will be no help on estate planning. And the guy down the street who knows a guy down the street is not to be trusted.

Believe it or not, everyone has an estate. An estate comprises everything, including cars, home, other real estate, checking and savings accounts, investments, life insurance, furniture, personal possessions, and in this specific case, telescope equipment. The old saying "You can't take it with you when you die" applies, even for beloved telescopes.

Estate planning enables control of how those things are given to the people or organizations that are important. To ensure your wishes are carried out, you need to provide instructions stating *whom* you want to receive something of yours, *what* you want them to receive, and *when* they are to receive it. You will, of course, want this to happen with the least amount paid in taxes, legal fees, and court costs. This is why a professional lawyer and CPA are integral to the estate planning.

And prior to or during any estate planning, find an astronomy expert to assess the telescope equipment and provide a professional appraisal. The guy down the street who has a telescope and buys from eBay doesn't count. Many telescope shops across the country can provide this service at minimal cost. Some local astronomy clubs may provide assistance in locating an individual who can provide actual, timely, and fair valuation to astronomical equipment. Who knows, you might be lucky to have PBS's Antiques Roadshow coming into your area to film a show!

Wills and Living Trusts—Astronomy Loving Families

There are two groups that will be discussed: (1) families who love and participate in astronomy and (2) families who have no interest in astronomy.

There are those who are fortunate to have sons and daughters who love astronomy as a hobby and are avid observers. Starting at an early age and introducing the art of observing the stars, galaxies, and planets through the eyepiece can lead a child towards the study of the hard sciences such as physics and chemistry in college and a possible career in the sciences or engineering. Wills and living trusts bequeathing telescopes becomes an easy task in this case, knowing that the offspring has a lifelong interest that began in childhood. It simply becomes a decision as to who gets what telescope or which eyepiece.

A listing of the telescope(s) and associated accessories can be made as an attachment to the will, with the wishes of who gets what. Lawyers find this method easier and cleaner than including the bequest directly within the will. And hopefully you can divide up the goodies fairly. Good Luck!

Another strategy is to begin giving your children some of the astronomical equipment over a period of time prior to your daisy pushing time. This strategy is applicable especially for those who have chosen the Living Trust route. This works especially well when larger heavier telescopes become too cumbersome as you age. And for those who worry about such things, the Federal Government allows annual gifts to offspring up to $10,000 with no tax penalty.

Wills and Living Trusts—Families Who Have No Interest in Astronomy

Unfortunately, there are many families where the astronomy passion never took hold.

That's okay. Although it's too bad they never discovered the love of astronomy, this diverse world holds many fascinations that can attract and hold someone's attention. However, that still leaves the disposition of the late astronomer's equipment as a major task for families.

The process begins in the same manner as before. Get an appraisal of the astronomical equipment. Seek out a professional lawyer, estate planner, or the assistance of a certified public accountant to develop a will or living trust. Once that is done, a family discussion is still needed to go over your wishes on the disposition of the equipment (along with any other items).

It is at this point that the process diverges from astronomy-loving families. And a point where your wishes must be clearly stated and documented by your estate professionals. It will then be the responsibility of the executor of your estate to execute your wishes.

Options for Disposition of Heirloom Grade Telescope Equipment

There are several options that must be explored when leaving the telescope equipment and astronomy books to:

1. A friend or young person who loves astronomy. During one's lifetime, one makes many friends within the astronomy community. A nice final gesture is to acknowledge these friends by bequeathing your optical collection to them.
2. An astronomy club. Every region of the country has a local astronomy club that is involved with public education and outreach and supporting young people entering the sciences. What better way to support these organizations than leaving your treasured telescope equipment for the benefit of the community. There are some astronomy clubs that may chose to distribute the telescope equipment among its members. If this is okay with you, this is a plausible option.
3. A school or education institution. As with astronomy clubs, supporting public education of the young is a fine use of your telescope. The only caveat to this option is finding a school science department that is educated in the telescope's use and is willing to develop a program or educational curriculum that takes advantage of this gift. Otherwise, as with many school systems around the country, telescopes will sit in a storage room and remain unused until the school surpluses them to a public auction to raise educational funds. This has happened in Frederick County, Virginia, where Walmart gifted two Meade 16″ SCT telescopes and eyepieces. The telescopes have sat unused and still in their crates, with the school system unable to find a teacher willing to develop and support an astronomy program.
4. Gradual sell-off of equipment. This option allows you to be in full control of the disposition of the telescope equipment, and the distribution of the funds following the sale. This option is viable over a set period of time, but does not take into account of sudden death.
5. Sell the equipment through an estate sale, auction or online auction, and spread the proceeds among the family survivors.

Selling Through an Estate Auction

This chapter is not intended to be a detailed primer on estate tag sales, estate auctions, online auctions, and the like. Again, seek out professional auction houses for detailed guidance. Just be aware of the various common options that are open:

1. Antique Auction—An auction offering antiques for sale from one or more individuals. For older and classic telescopes, this can be used as an auction option. This type of auction appeals to two types of bidders, those seeking telescope equipment for astronomical use and those interested in antique scientific instruments.

2. Estate Auction—An auction offering the contents of a single individual's estate. While this type of auction has traditionally involved items from a deceased individual's home or collection, modern usage has evolved into the term "Living Estate" which, of course, means that the owner is still alive. This type of auction draws all types of clientele, many not interested in paying top dollar for telescope equipment. Many attendees of this form of auction can be "bottom feeders" only interested in bargains and flipping items for profit.
3. Estate Tag Sales—These are a subset of estate sales where every item is tagged with a price, and as the tag sale continues over many days, each day brings a lower tag price for each item. As with the estate auction, this option draws all types of clientele, many not interested in paying top dollar for telescope equipment. Again, many attendees of estate tag sales can be "bottom feeders" only interested in bargains and flipping items for profit.
4. Consignment Auction—An auction offering items from multiple owners, and a mix of items of undetermined age. This type of auction draws all types of bidders, many not interested in paying top dollar for telescope equipment. As before, many attendees of this form of auction can be "bottom feeders" only interested in bargains and flipping items for profit.
5. Unreserved Auction—An auction offering goods to be sold to the highest bidder, regardless of price. Many galleries and auction houses have minimum or "Reserved" bids, which set a minimum starting price for each item. These auctions can have different legal implications depending on the laws of the State in which the auction is held.
6. Scientific Instruments Auction—As seen in Figs. 11.2 and 11.3, there are occasionally auctions limited to just scientific instruments. These types of auctions are rare, but will represent and highlight the telescope and its accessories to yield the highest price and greatest return. Figures 11.2 and 11.3 are from an upstate New York scientific instrument auction held in the early 1990s. Classic Clark, Mogey, and Brashear refractors were sold for high prices, along with more contemporary Zeiss and Brandon refractors.

The downside to any auction is the auction house's piece of the action. Often, the auction house will retain 10% of the total price as a seller's fee. Beware, there are some auction houses that will ask for as much as a 30% fee.

Selling Through eBay Auctions

For those who have been hiding under a rock for over a decade or two, eBay is an online auction service used to buy and sell items. The company uses an electronic platform to facilitate millions of transactions every day. Users seeking to purchase items make bids over a specific time period and then the seller determines guidelines such as a minimum bid the seller is willing to accept. Payment is by an online payment system known as PayPal, or by credit card or snail-mail check.

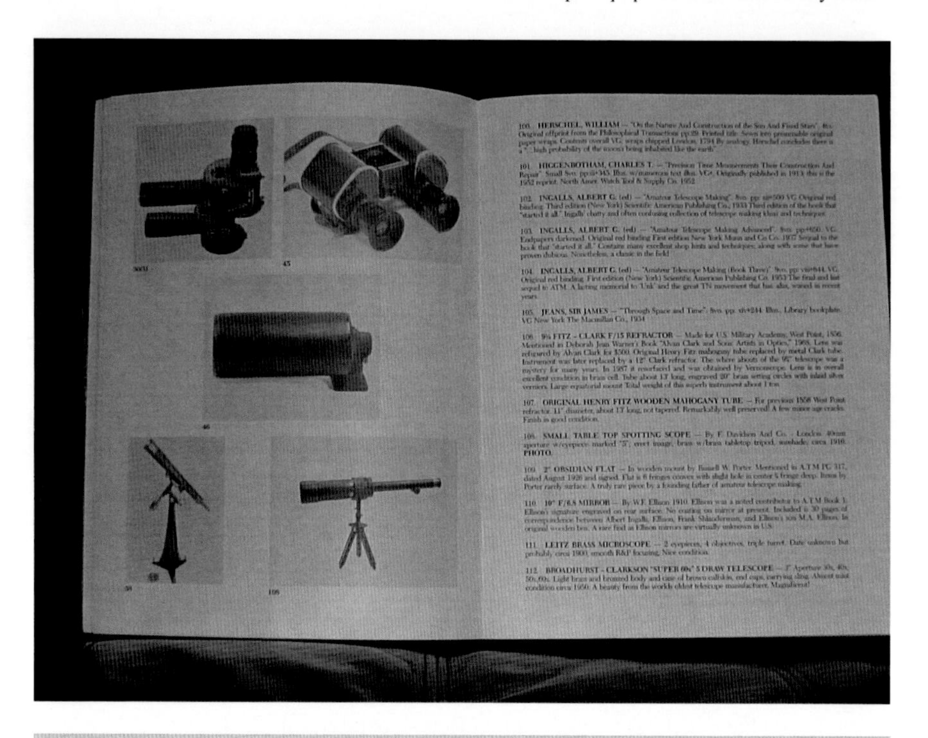

100. HERSCHEL, WILLIAM —

101. HIGGENBOTHAM, CHARLES T. —

102. INGALLS, ALBERT G. (ed) —

103. INGALLS, ALBERT G. (ed) —

104. INGALLS, ALBERT G. (ed) —

105. JEANS, SIR JAMES —

106. 9⅝ FITZ - CLARK F/15 REFRACTOR —

107. ORIGINAL HENRY FITZ WOODEN MAHOGANY TUBE —

108. SMALL TABLE TOP SPOTTING SCOPE —

PHOTO

109. 2" OBSIDIAN FLAT —

110. 10" F/6.8 MIRROR —

111. LEITZ BRASS MICROSCOPE —

112. BROADHURST - CLARKSON "SUPER 60s" 5 DRAW TELESCOPE —

Fig. 11.2 A selected page from the 1989 Vernonscope Scientific Instruments Auction

Fig. 11.3 Another selected page from the 1989 Vernonscope Scientific Instruments Auction

If you know what eBay is, or have used eBay for buying and selling, good for you. If you don't know what eBay is or what to do, there are computer-literate friends, books, and online tutorials that are available to help you. This part of the chapter is not meant as a tutorial on how to use eBay.

For your survivors wanting to dispose of your precious telescope equipment, selling it on eBay is a somewhat labor intensive approach that has a high probability of yielding high returns. Millions of people shop eBay daily, looking specifically for certain categories of items and products or very specific items and placing bids on these items. The winning bid results in a payment by the buyer to the seller, but also means the seller must deliver, in person or by UPS, FedEx, DHL, or USPS, the item as described.

Selling on eBay can be labor intensive. As tempting as it is to sell the whole telescope collection as a single lot, this strategy is for the lazy and for those who don't want to maximize their return.

Here are the best tips for maximizing returns on eBay auctions:

1. List each telescope to be sold separately. List each telescope appropriately so the eBay search engine can find it. For example, if selling a Stellarvue 80 mm refractor telescope on an alt-az mount, list it as the Stellarvue 80 mm refractor telescope on an alt-az mount. By failing to give the brand name (in this case Stellarvue), those potential bidders searching for Stellarvue will miss seeing the auction.
2. List each eyepiece to be sold separately. Or if a complete set of a brand name or design eyepieces. For example, if a complete set of Brandon eyepieces are up for auction, each separate eyepiece can represent a single auction, or the whole set can be auctioned. If a mix of eyepieces exists of different brands and designs, it is best to list each eyepiece separately.
3. Establish a reasonable starting bid and reasonable reserve. Price things too high and no one will bid on it. Check for similar items being sold on eBay as a guide for the starting price, reserve, and incremental minimal bid levels.
4. Set a reasonable incremental bid increase.
5. Shipping should be part of the buyer's cost, unless the item is small (like an eyepiece or filter). Free shipping is a draw to bidders.
6. High-quality pictures should accompany each item for auction.
7. Describe each item in detail, both condition of the item and the original manufacturer's description of performance and design. Like-new or new-in-box condition brings top dollar. Be honest in the auction listing. If there is a screw mark on an eyepiece barrel, note it in the auction description. This will save you from potential eBay idiots (see number 11 below).
8. Use accurate terminology when writing and describing the auction.
9. A 2-week auction is sufficient. Too short an auction period means you miss potential bidders. An auction period that extends for too many days or weeks will cause potential bidders to forget about the auction. Most of the bidding tends to occur during the last hours of the auction, with a great deal of competition happening during the last minutes and seconds.

10. Timing of the auction during the year is critical to maximizing auction prices and bidding activity. Avoid the summer season as everyone is focused on vacations and are not interested in eBay auctions. Late fall and the Christmas rush have the greatest yields. Late winter and spring are also active times on eBay as people receive their tax returns and are eager to spend money.
11. Ship the item to the winner of the auction as soon as payment has been verifiably received.
12. Be prepared to deal with returns and refunds, or buyers who have unreasonable expectations. It comes with the territory. Some people buying on eBay are idiots.
13. Keep an accurate accounting book of eBay auctions, listing costs, eBay fees, asking price, selling price, and total return. This helps when dividing up and accounting for all of the items.
14. Keep an accurate record of all transactions for tax purposes.

To all of your heirs, tell them what has been sold after all the auctions and eBay sales, but don't be surprised if there are things leftover unsold. At your discretion, you can pass on your instructions on what to do with the leftovers (let's keep it clean here!).

Chapter 12

Final Thoughts

Astronomy, as
nothing else can do
teaches men humility.

—Arthur C. Clarke

Astronomy has impacted the lives of civilizations, cultures, and individuals since the dawn of mankind. People from all ages and all regions of Earth have stared into the night sky with awe and wonder.

The stars and galaxies were there before we were born.

The stars and the galaxies will be there long after we pass through our individual lifetimes.

As you get older, it becomes important for astronomy lovers to pass their knowledge about the universe to the younger generation.

We grow older, and our families expand to include children and grandchildren. The amateur astronomy community needs to pass our love and passion for knowledge and the appreciation of the beauty of the night sky to the next generation. Nothing is more rewarding for a father or mother to share time at a telescope with our sons and daughters. We as grandparents can find great joy with grandchildren as they discover the night sky together. Astronomy is good family time.

Astronomy is woven into the fabric of culture. Knowledge of astronomy helps our children and grandchildren to more fully comprehend and appreciate classical literature and its references to the stars. From ancient texts of Plato to popular science fiction to current bestsellers, astronomy often plays a role in plot lines and character development.

J.L. Chen, *Astronomy for Older Eyes*, The Patrick Moore Practical Astronomy Series, DOI 10.1007/978-3-319-52413-9_12

Fig. 12.1 Vincent Van Gogh's The Starry Night (van Gogh)

Some of the greatest minds in classical and western literature wrote on the subject of astronomy: Homer, Virgil, Aristotle, Cicero, Ptolemy, Euclid, Chaucer, Leonardo da Vinci, Shakespeare, Jules Verne, H.G. Wells, James Michener, Arthur C. Clarke, and many others. This is because astronomy was part of their daily lives.

Some of the greatest artists have paid homage to the stars. Think of Vincent Van Gogh's *Starry Night* (Fig. 12.1).

Apollo 12 astronaut Alan Bean has become a painter, often mixing particles of Moon dust into his works (Figs. 12.2 and 12.3).

Serious astronomy has been accepted on a popular cultural level, with the recent popularity of such movie box office hits as *Gravity, Interstellar,* and *The Martian* (Fig. 12.4). The mythic *Star Wars* and the long running *Star Trek* series of movies and television shows have become part of our popular culture and even part of our lexicon. Travel to other stars and solar systems has become so common place in the movies that some of the general public thinks it is real. How sad and disappointed they are when fiction is not reality.

Astronomy is represented in music, from Gustav Holst's *The Planets*, Pink Floyd's *Eclipse* from the *Dark Side of the Moon* album, to the theme song for the television hit *The Big Bang Theory.*

Astronomy helps us understand the rhythms of our daily lives. The Earth turns on its axis to produce our days and nights. The Earth orbits the Sun and nature responds with seasonal cycles culminating with our understanding of a calendar year.

Fig. 12.2 Moon painting entitled "America's Team...Just the Beginning" by Astronaut Alan Bean (Alan Bean)

Fig. 12.3 Moon painting entitled "Home Sweet Home" by Astronaut Alan Bean (Alan Bean)

Astronomy and mathematics are integral with each other. The scientifically astute recognize how higher mathematics combine with physics to describe astronomy and nature in ways that are still being discovered. The ancient craft of navigation and timekeeping are based on the mathematics of the basic motions of the heavens. Isaac Newton's *Principia* described how the world works with mathematical explanations. Albert Einstein's *General Theory of Relativity* and *Special Theory of Relativity* play a part in everyday lives, often unbeknownst to the common individual. Without Einstein's equations, satellite navigation, commonly referred to as

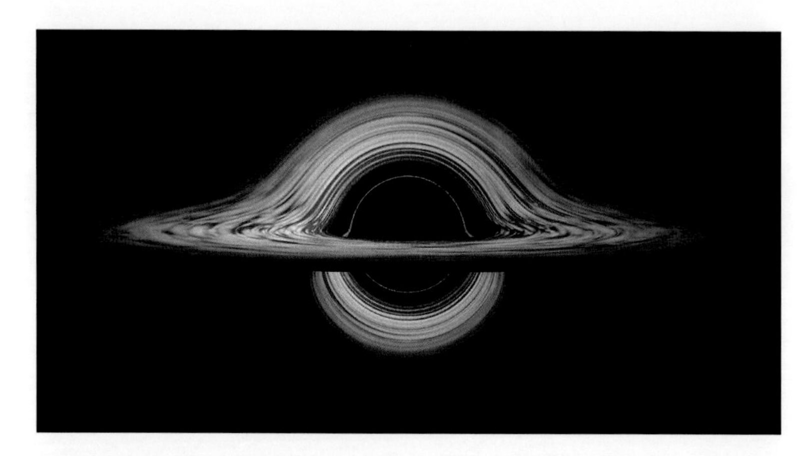

Fig. 12.4 NASA simulation of a black hole accretion disk, including the light bending effects of relativity. Similar to black hole depicted in the motion picture **Interstellar** (NASA/Jeremy Schnittmann)

GPS or SATNAV, wouldn't work. The history of mathematics is inextricably bound to the history of astronomical studies.

Astronomy should be part of a well-rounded education. Astronomy used to be a part of the time-honored seven Liberal Arts of classical education. By studying astronomy, students become a part of a long-standing tradition that has benefited western learners for hundreds of years.

Astronomy broadens a person's outlook on the world and the universe.

Astronomy helps everyone to view life on a grander scale.

Astronomy stimulates the mind, and the stimulation of the older mind helps extend mental sharpness well into old age.

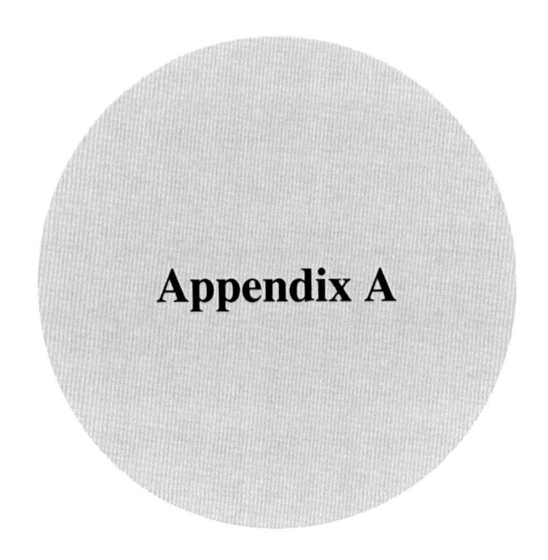

Appendix A

Telescope Basics

The Telescopes

The world of astronomy is inhabited by a menagerie of telescopes. There are short-focus and long-focus refractors, Dobsonian reflectors, Schmidt– or Maksutov–Cassegrains, Newtonian reflectors, GoTo telescopes, achromats, apochromats, … the list goes on and on. There are telescopes of every size and for every budget. Some with manual altitude-azimuth mounts, some with German equatorial mounts, and some with very sophisticated electronic GoTo mountings. It's no wonder that a person new to astronomy gets confused and intimidated (Fig. A1).

The following discussions on telescope types demand a definition of focal ratio, or f/ratio. Quite simply, the *f*/ratio is the focal length of the telescope divided by the diameter of the main lens or mirror. The smaller the *f*/ratio, the lower the magnification and the wider the field of view with any specific eyepiece. Higher magnification is easily attained with a higher *f*/ratio, but with the cost of a smaller field of view.

In order to simplify the world view of amateur astronomy, it is best to organize telescopes into the three basic categories: refractors, reflectors, and catadioptrics.

J.L. Chen, *Astronomy for Older Eyes*, The Patrick Moore Practical Astronomy Series, DOI 10.1007/978-3-319-52413-9

Fig. A1 Group photo of the author's collection of telescopes (James Chen)

The Refractors

Ask someone to close their eyes and picture in their mind a telescope. Or better yet, ask a child to take a crayon and draw on a piece of paper a picture of a telescope. More likely than not, the image of a long tube pointed at the skies with an observer peering through the opposite end is the result. The refractor is the intuitive concept of a telescope.

Historically, the refractor is the earliest design for telescopes, with the earliest examples appearing in the Netherlands in 1608. Spectacle makers Hans Lippershey and Zacharias Janssen are two of the early creators of the design. Within a year or two Galileo created his own improved refractor design, pointed his telescope into the night sky, and history was made.

Conceptually, a refractor is a system of lenses with an objective lens system to gather light and an eye lens system to focus the light gathered by the objective lens into an image for the observer. Hence, the intuitive mental picture of a long tube with an objective lens pointed at the sky and an observer viewing through the opposite end focusing the eyepiece.

Inch-for-inch, refractors are regarded by the astronomy community as the best performing telescopes. Sharp, pinpoint star images, and high-contrast planetary images are their hallmark. Four inch or larger diameter refractors are the primary instruments of lunar and planetary observers. Astrophotographers have adopted apochromatic refractors as their go-to telescopes, producing the sharp and contrast images that rival professional observatories.

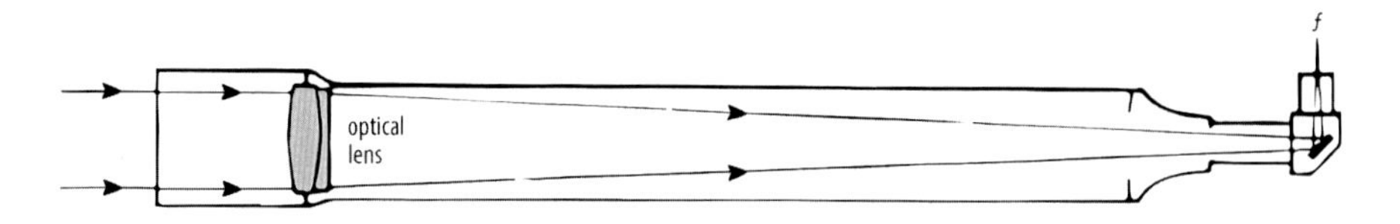

Fig. A2 Classic Doublet Achromatic Refractor (Adam Chen)

In today's world of refractor telescopes, there are achromats and apochromats, terms created to differentiate the levels of color correction of the respective lens systems. Grade school science demonstrates that sunlight passing through a prism separates sunlight into its constituent colors. In optical systems, such as telescopes, all types of optical glasses exhibit some degree of separation and dispersion of light into primary and secondary colors (Fig. A2).

A classic achromatic refractor uses a two lens objective, with one lens made of crown glass and the other lens made of flint glass. With the lenses ground with proper curves and using glasses with different refractive indices, the result is a telescope that can bring to focus two of the three prime colors of light, typically red and blue wavelengths. So conceptually, the achromat can produce an improved image of a distant object over a single lens objective of the type used by Galileo. In practice, there are some color errors, or chromatic aberrations, that creep into the visual performance of an achromatic refractor, especially at shorter focal lengths. The primary colors do not focus at the same point, resulting in a color fringe around Moon, planets, or bright stars. This chromatic aberration also results in diminished sharpness and definition in the telescope image. Many classic early refractors typically have focal ratios of *f*/11, *f*/15, or *f*/20 or greater to minimize chromatic aberrations and become "color-free." Even at long focal lengths, achromats can display chromatic aberrations, where the secondary colors of yellow and purple wavelengths do not come to focus.

This technical discussion gives the impression that achromats are flawed. That couldn't be further from the truth. Today's achromatic designs use different types and combinations of Extra-low Dispersion, or ED, glass to lower the chromatic aberrations to a minimum. Modern designs of achromats have minimized the false color to a high degree. One current builder of high-quality refractors has taken advantage of new glass technology and our light polluted skies. With a combination of ED glasses, altered lens curves, and adjusting the air spacing between the two lenses of the doublet, the color error in the violet wavelengths matches the not-quite-black-really deep-purple light polluted background sky, resulting in an excellent performing telescope that produces sharp and crisp images with the color error conveniently and cleverly hidden in the background. There are specialized long-focus 102 mm achromatic refractors on the market with *f*/11 focal ratios or greater specifically designed for viewing the Moon and the planets. These modern "planet killers" take advantage of low dispersion glasses and long focal length to provide images that rival apochromatic refractors at one-third the cost. In the case of viewing deep sky objects, there are affordable achromats, up to 6 in. in aperture, that

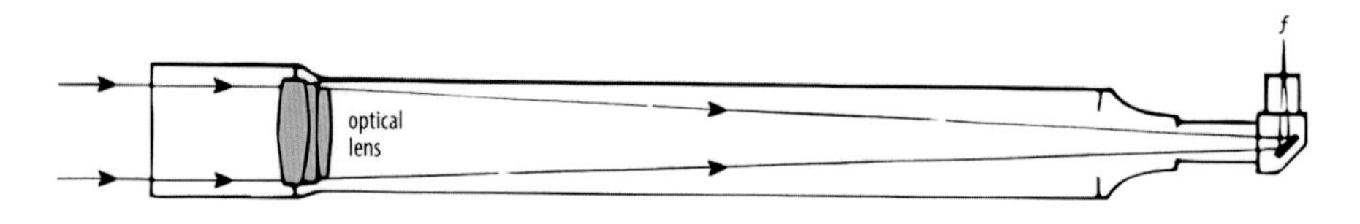

Fig. A3 Modern Triplet Apochromatic Refractor (Adam Chen)

offer the refractor's higher contrast than reflecting designs, and any chromatic aberration effects are less detectable on the dimmer deep sky objects (Fig. A3).

Sometime in the 1980s, the apochromatic refractor became commercially available to the amateur astronomy market. Sophisticated designs appeared using combinations of two, three, or four lenses of exotic glasses. Advanced telescope owners found themselves developing and reveling in a new vocabulary that included terms such as fluoro-crown, lanthanum, fluorite, FPL-51 or -53, APO-triplet, and Petzval. These apochromats offered improved sharpness, contrast, and color correction over the classic achromat refractors, while at the same time creating a more portable telescope with f-ratios of f/8 or less. But this optical performance improvement is costly. Apochromats are typically 2 to 10 times more expensive than an equivalent aperture achromatic refractor. Apochromats get expensive very quickly as the size increases. A typical 80 mm apochromat retails in the $700 range. A 102 mm triplet can easily cost over $2500. One hundred thirty millimeters apochromats range falls into the $4000–$5000 realm, and you can buy a car with the money needed for larger apochromats.

The old telescope salesman explanation of an achromat refractor is a "color-free image," meaning no color fringing in the telescopic image. The telescope salesman pitch for an apochromat is "this time, I'm serious! It's color-free!". Depending on the sensitivity and sensibility of the observer, the image improvement of an apochromat over an achromat can be either a slight or a day-night difference. A lot depends on the sensitivity and sensibility of the refractor owner. From a technical viewpoint, there is no question that the apochromatic refractor offers the best telescopic images over an achromatic refractor. In fact, as discussed later in this chapter, an argument can be made that an apochromat is inch-for-inch better than every other telescope design, whether refracting, reflecting, or catadioptric.

The Reflectors

Countless astronomy books and websites over the years have presented valid arguments that the Newtonian reflector represents the most "bang for the buck" in telescopes. Originally designed by Sir Isaac Newton in 1668, this telescope represents a simple design using either a spherical or parabolic main mirror to collect light and a flat diagonal mirror to reflect the collected light to the eyepiece and the observer. By limiting the number of optical surfaces that need to be accurately figured and

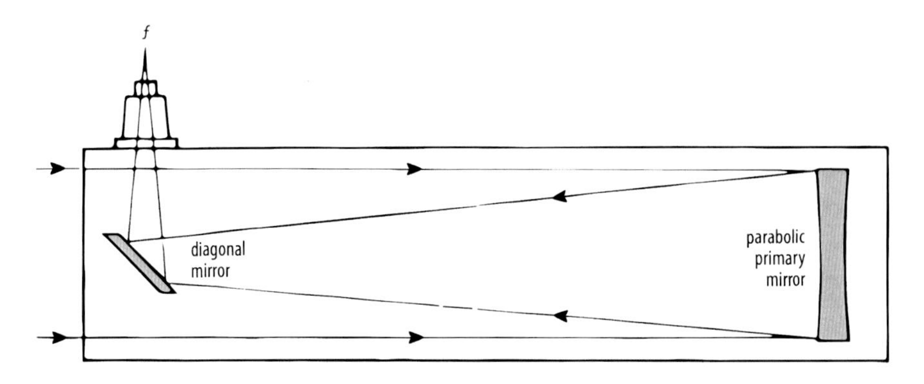

Fig. A4 The Newtonian Reflector (Adam Chen)

polished to two, the cost for a Newtonian telescope can be minimized. In contrast, a doublet refractor needs four optical surfaces, and a triplet apochromat needs six optical surfaces to be accurately ground, figured, and polished. The Newtonian design, being reflective in nature, has no chromatic aberrations and can be constructed from less expensive materials. During the explosive growth of Newtonian telescopes in the 1980s, with John Dobson's unique implementation of the design, Newtonians could be made from Pyrex glass, cardboard tubes, and plywood (Fig. A4).

Sounds like the perfect telescope. An 8-in. Newtonian reflector on a Dobsonian mount can be easily acquired for the price of a 3-in. achromat refractor. That is a lot of performance for the money.

But there is a catch. Ask any telescope dealer, any telescope salesman, or any member of your local astronomy club, "What is the most common telescope problem?" Nine times out of ten, owners of Newtonian telescopes are unable to align the optics of their telescopes. The main mirror is tilted, the diagonal is not centered, the tilt of the diagonal needs adjustment, or a combination of all three leads to frustration for the telescope owner, resulting in a telescope that ends up in the closet. Whereas refractors and some catadioptric designs do not need collimation alignment under normal use, Newtonian telescopes require frequent maintenance. Transporting the Newtonian by car or just the act of moving the telescope for inside a home to the patio can jar the Newtonian's optics to misalignment. Owners of Newtonians must acquire the skill for realignment of their telescopes. This alignment skill is easily acquired by some, while others find the alignment task difficult. Some telescope owners take collimation in stride, while others find it a nuisance.

The Catadioptrics

Catadioptric telescopes are a category of telescopes that combine lens and mirror technology to produce compact and transportable instruments. With a clever combination of a lens and mirrors, the incoming light path is folded upon itself, and any

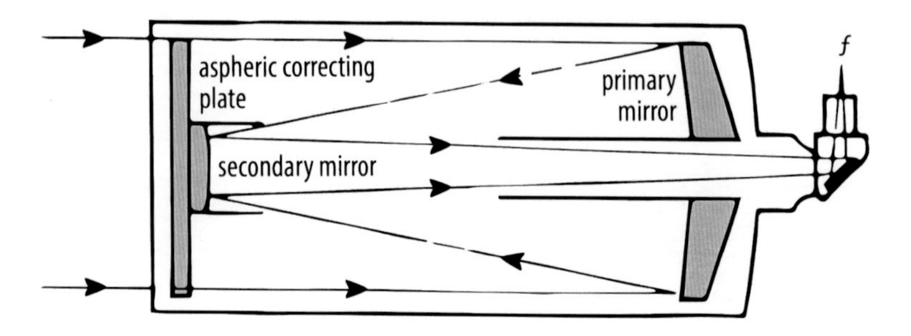

Fig. A5 The Schmidt–Cassegrain Telescope (SCT) (Adam Chen)

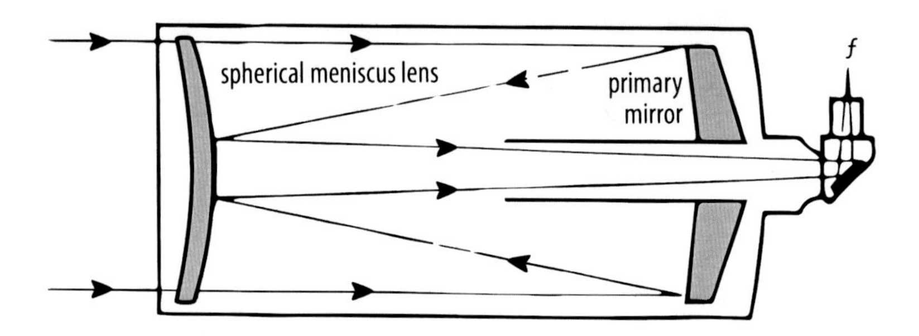

Fig. A6 The Maksutov–Cassegrain Telescope (Adam Chen)

optical aberrations of the reflecting surfaces can be corrected by the refracting lens (Figs. A5 and A6).

Catadioptrics are available in two popular forms: Schmidt–Cassegrain telescopes, or SCT, and the Maksutov–Cassegrain telescope. Maksutov–Cassegrains appeared on the commercial market in the mid-1950s, and SCTs burst into popularity in the 1970s. As a family, catadioptrics offer a high degree of optical performance and low maintenance in a small compact package.

Although SCTs are available in sizes from 5-in. to massive 14- and 16-in. apertures, the most popular and best-selling telescope since the 1970s has been the 8-in. SCT. The 8-in. SCT is the basic core product for not one, but two major telescope companies. The reason for this popularity stems from the all-around versatility that the design provides. As an analogy, the 8-in. SCT is the Olympic Decathlon Champion of telescopes. The Olympic Decathlon gold medalist doesn't excel in any singular competition, such as the 100 m dash or the high jump, but is able to perform ten different events better than anyone else can perform those same events. Versatility and a high level of performance are outstanding attributes of an 8-in. SCT. The 8-in. aperture allows for light gathering for deep sky objects, the typical *f*/10 focal length enables high magnification for lunar and planetary observations, numerous attachments are available for photographic use, and the compact size ensures frequent usage. But it's not perfect for any singular use. Newtonian telescopes are available at a cheaper price in larger sizes for deep sky light gathering. Refractors produce images that are of better quality in contrast and sharpness. Many amateur astronomers tend to own two, three, or more telescopes in order to optimize their viewing. But if a person is only going to own one single telescope that is capable of handling multiple astronomy tasks, the 8-in. SCT is the likely choice.

Maksutov–Cassegrain, or Maks for short, are highly popular smaller telescopes in the 3.5-to-6 in. diameter range. Costing from $400 to an astronomical (pardon the pun!) $4000, these little gems of the telescope world are available in every configuration conceivable. Maks tend to have long focal lengths in the *f*/12 to *f*/15 range, which helps to improve contrast but limits the field of view. Available on simple manual mounts, to equatorial mountings, to high-tech computerized go-to systems, Maks are as versatile as their SCT counterparts. Due to the compact size, Maks make excellent travel scopes.

The natural question that arises from all these discussions of telescope design is: What telescope is ideal? Recalling the car analogy, it depends on the type of observing.

For observing the Moon and planets, the answer is clear. Refractors offer the sharpest, clearest, and most contrasty images. But why? A discussion is needed on the difference between unobstructed and obstructed optics. As seen in Figs. 2.4, 2.5, and 2.6, the Newtonian and the catadioptric designs share a common attribute of a secondary mirror that is centrally located along the light path. These secondary mirrors allow for the folding of the light path to direct the incoming light to the eyepiece. In the case of the Newtonian, the secondary diagonal serves to guide the light to the side of the telescope to the focuser. In both catadioptric designs, the secondary mirror folds the light path creating a very compact telescope that enables portability. The commercially available Newtonians have secondary mirrors that measure approximately 30–35% the diameter of the primary mirror. In most SCTs and Maks, the central obstruction of the secondary mirror constitutes 40% the diameter of the primary mirror. The main impact of the central obstruction is a decrease in sharpness and contrast in the telescope image. What is the difference in performance between the unobstructed refractor and an obstructed design? The best analogy is to compare the familiar standard definition TV versus HDTV. A good SCT or short-focus Newtonian will offer a good lunar image like that of a standard 480i digital TV image, but a good ED achromat or apochromatic refractor is like watching a 1080p high-definition TV.

Any of these telescope designs can be used to view the HST objects. If the reader of this book currently owns a telescope, proceed to the following chapters and find the HST objects and enjoy. To those looking to purchase a telescope, the following recommendations are offered based on telescope technology, decades of amateur astronomy field experiences, and telescope retail experience:

1. For lunar and planetary observing, an 80 mm-to-102 mm refractor is the top choice. These refractors are light, portable, and low maintenance telescopes that offer sharp contrasty images of the Moon and planets. Remember, the reason that large telescopes exist is to act as light buckets and gather the faint distant light of

stars, galaxies, and nebulas. The Moon is bright. Really bright. Light gathering is the least of your problems in lunar observing. The major planets are bright, too. Larger apertures do help with planetary viewing when the observer is seeking greater detail. When aperture fever occurs, and a larger bulkier telescope is acquired, the refractor takes the role of the easy-to-use, grab-and-go scope. No wasted money here.

2. As discussed earlier, it's hard to argue against an 8-in. SCT. Yes, the lunar and planetary images are a little soft compared to a 102 mm apo refractor, but the overall versatility cannot be denied.
3. A long-focus *f*/10 or greater Newtonian has a central obstruction approximately 25% of the primary diameter, and comes very close to refractor image quality. These telescopes are difficult to find in the commercial market, but there are some of these gems on the used market.
4. 90–125 mm Maksutov–Cassegrains should also be considered, especially when small size is a requirement. However, these telescope will be aperture and wide-field challenged when viewing deep sky objects.

The Eyepieces

Telescope design is only half of the optical story. The rest of the story are the eyepieces at the focus end of the telescope (Fig. A7). And here again, there's another zoo filled with strange and wonderful denizens that are vying to complete the

Fig. A7 The author's collection of telescope eyepieces (James Chen)

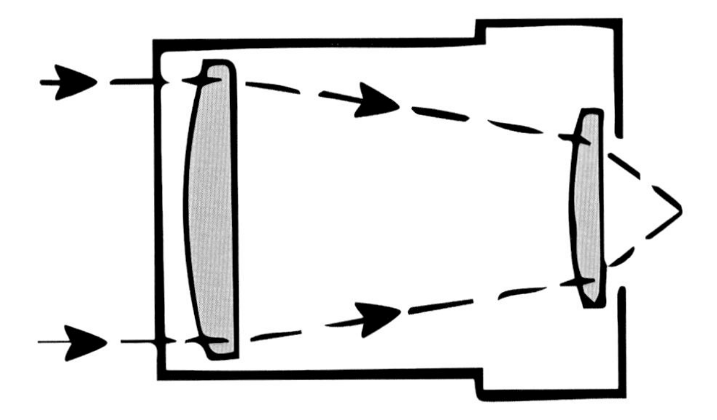

Fig. A8a Huygens (Adam Chen)

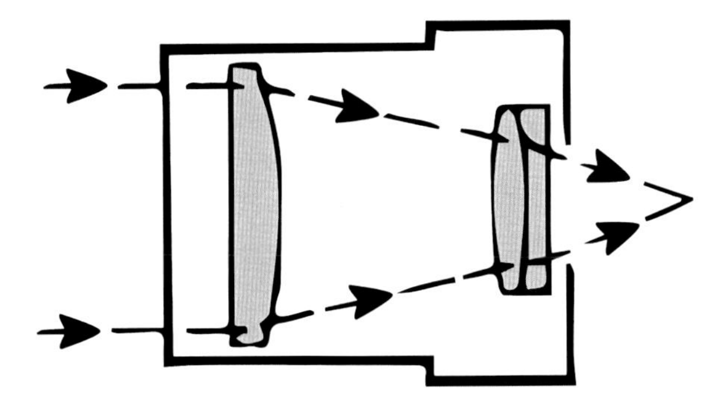

Fig. A8b Ramsden (Adam Chen)

optical train. Kellner, Abbe, Konig, Brandon, and Nagler are all names of optical designers who have lent their names to their eyepiece designs, and are now an accepted part of the astronomy vocabulary. There are 0.965-, 1.25-, and 2-in. eyepieces. Some designs have been around for over a century, and some designs are less than a decade old. And most have valid use in astronomy (Figs. A8a and A8b).

Two of the oldest and simplest of the compound eyepiece designs are the Ramsden and Huygens. Originating from the 1700s, these eyepieces serve as historic curiosities. Occasionally, an antique Ramsden will show up at swap meets, eBay, or even antique stores. The Huygens eyepieces are still supplied in 0.965 in. size on cheap beginner telescopes sold at department stores and big-box stores. Both designs are flawed, with narrow apparent fields, chromatic aberration, and short eye relief (Fig. A9).

In the mid-1800s, the Kellner eyepiece was developed by replacing the singlet eye lens element of a Ramsden with an achromat doublet. This resulted in a better performing design with a wider field, better color correction, and less spherical

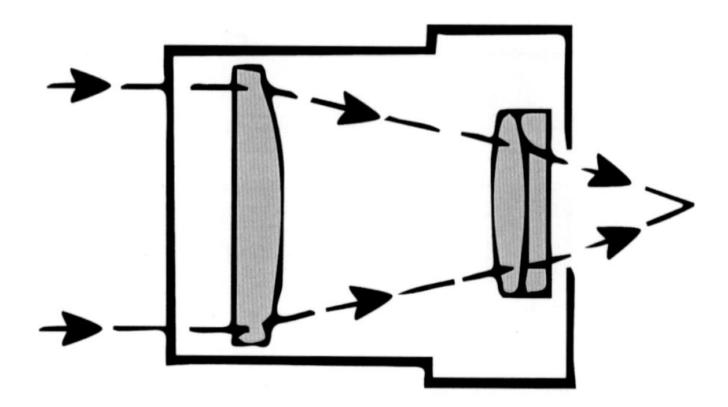

Fig. A9 Kellner design (Adam Chen)

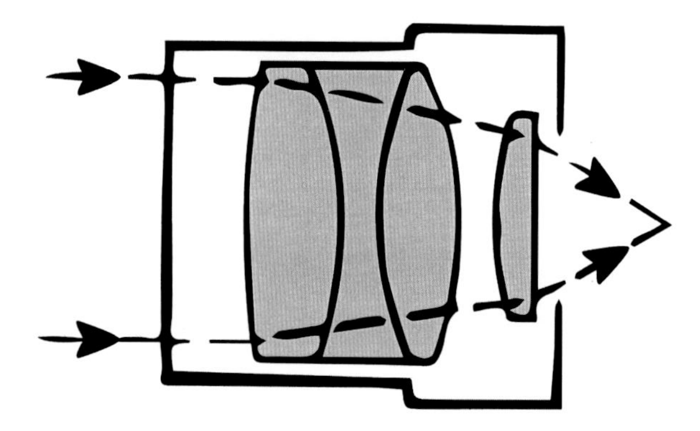

Fig. A10 The Abbe orthoscopic (Adam Chen)

aberrations. When used on long-focus telescopes, Kellners still produce a reasonably good image. The main shortcoming of the Kellner is ghosting when looking at bright objects. So Moon watchers beware! The design exists today under various names, including modified achromat, RKE, and modified Kellner. These eyepieces are a good economical alternative for those on a budget (Figs. A10, A11, and A12).

An examination of the patents for the Abbe, Plossl, and Konig eyepiece designs describes all three as orthoscopic designs. The term "orthoscopic" means free from distortion. In common astronomy vernacular, the term orthoscopic has evolved to become synonymous in name with the Abbe design.

The Abbe has stood the test of time. Since the 1950s, through the growth of amateur astronomy in the 1960s and 1970s, the Abbe design has been highly regarded for sharp and high-contrast images. The classic "volcano top" Abbe orthoscopic, so named for its distinctive beveled shape, are well known and are highly desired. These Abbe eyepieces have been made by an optician from Japan named Tani-san, whose retirement in 2013 brought an end to decades of Circle T volcano

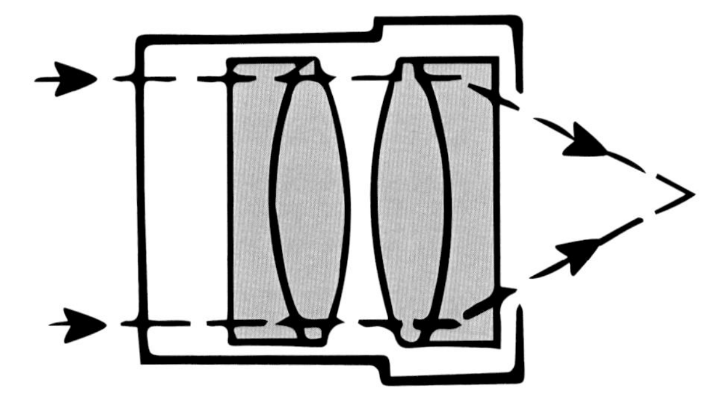

Fig. A11 The Plossl orthoscopic (Adam Chen)

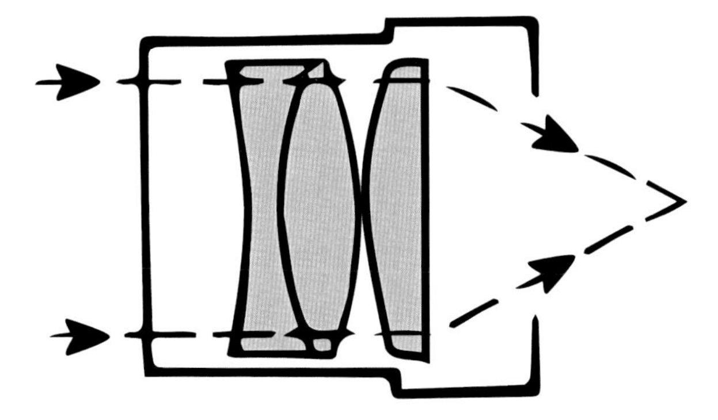

Fig. A12 The Konig orthoscopic (Adam Chen)

top eyepieces. But don't fret, Abbe orthoscopic eyepieces are available through other sources. The Abbe orthoscopic eyepiece was held up as the pinnacle of eyepieces until the advent of new revolutionary wide-angle and high eye relief eyepieces in the 1980s. Today, dedicated lunar and planetary observers still insist on Abbe eyepieces today.

The Plossl eyepiece has an interesting reputation in the amateur astronomy world. In the 1960s, the Plossl only existed as a rare and mysterious eyepiece, commercially available from a small vendor in Europe. Then in the 1980s, Plossls suddenly became widely available, to the point now where it is so commonplace that the eyepiece is considered mediocre by average backyard astronomer. Nothing could be further from the truth. Although poorly manufactured examples exist, a premium Plossl is an extraordinary eyepiece, versatile in lunar, planetary, and deep sky use. There exists a number of variants of the design in which an extra lens or two are added to the system, somewhat blurring the definition of a Plossl, but these variants tend to be of high quality and offer high performance.

Not as widely available as the other orthoscopics, the Konig design has its fans. Depending on the implementation, the Konig potentially offers a wider field of view than the Abbe or Plossl. In practice, the Konig achieves a wider field, but with a slight sacrifice of edge of field sharpness, and slightly shorter eye relief.

A variant of the Konig design is the Brandon eyepiece. Chester Brandon developed his eyepiece design during his time at the Frankford Arsenal in Philadelphia. The eyepiece design was widely used during World War II in U.S. Army optics. The Brandon design differs from the Konig by using three high index glass types and four different lens radii. When marketed in the 1950s as a high-priced premium eyepiece, the Brandons were priced at $15.95. My how times have changed, with the price of Brandons and many premium eyepieces in the three-digit range. With sharp crisp contrasty images, the Brandon reputation exceeded that of any other 1950s eyepiece, and is still the go-to eyepiece for many lunar and planetary observers today (Figs. A13a and A13b).

During World War II, a number of scopes used for spotting or for aiming weaponry by the American armed forces were equipped with Erfle eyepieces. In the

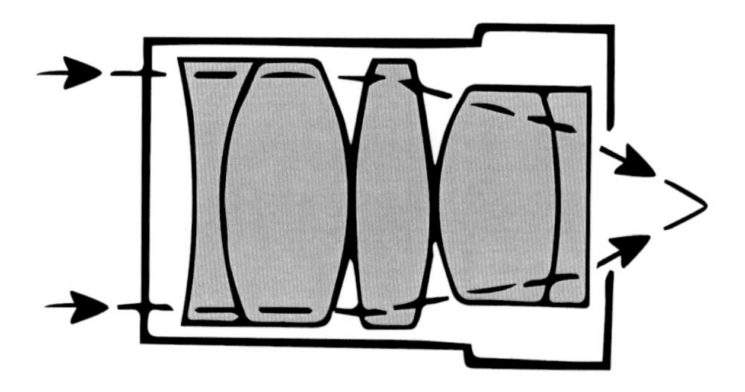

Fig. A13a The Erfle (Adam Chen)

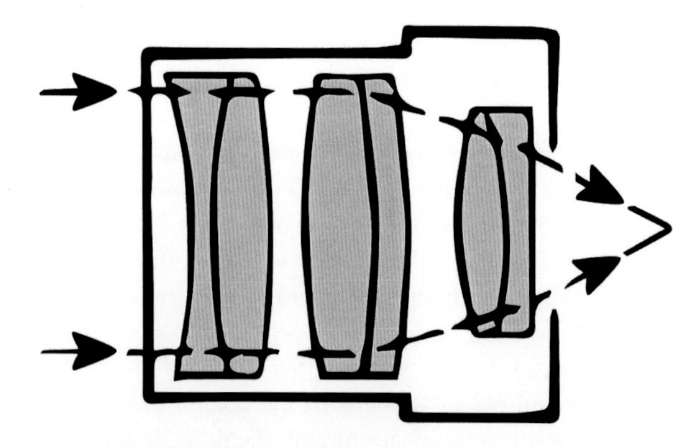

Fig. A13b Modern Wide-Angle (Adam Chen)

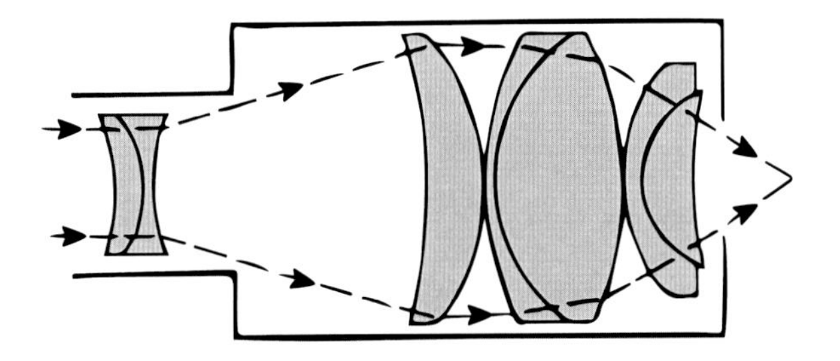

Fig. A14 The Nagler and related ultra wide angle eyepieces (Adam Chen)

1950s and 1960s, war surplus Erfle eyepieces became available for amateur astronomers seeking wide-angle views through their telescopes. With an apparent field of view of approximately 60°, these large eyepieces created a demand for wide-field eyepieces. That wide-angle demand has grown over the years, to the point now where the consumer demand for ever wider fields of view drives the eyepiece industry. The Erfle does not display as sharp an image in the center of the field as the orthoscopic eyepiece, and a degradation of the image occurs in the outer third of field. Newer modern wide-field designs use exotic glass types and different lens configuration and curves to correct the edge degradation, while at the same time providing a wider field of view (Fig. A14).

The quest for an ever wider field of view exploded with the introduction of the Nagler eyepiece. Competitors quickly followed, with virtually every telescope company offering their version of an over 80° field-of-view eyepiece. The ante was raised again with the introduction 100° and even wider field-of-view eyepieces in recent years. These eyepieces contain seven or more lenses in their complex designs in an effort to provide wide fields without sacrificing sharpness at the edge of the field. These eyepieces are not cheap, with many exceeding the cost of many telescopes! These eyepieces are outstanding for deep sky and wide-field applications. However the complex design, high number of optical surfaces, plus the inevitable, although slight, light absorption caused by the amount of glass in the light path, these ultra wide angle eyepieces do not offer the same level of sharpness and contrast for planetary and lunar observing as the simpler orthoscopic designs (Fig. A15).

While not an eyepiece, a Barlow lens is a useful addition to every eyepiece case. A Barlow lens is a negative lens system placed along the light path between the objective and the eyepiece that increases the effective focal length of the telescope, therefore increasing the magnification. Typically, Barlows double (2×) or triple (3×) the magnification. Newer focal extenders using three or four lens elements are available to quadruple (4×) or quintuple (5×) the focal length. These accessories are useful in three ways. A single Barlow lens effectively doubles the number of magnifications available in an eyepiece collection. The use of a focal extender also allows longer focal length eyepieces with their higher eye relief to be used at higher magnifications for eyeglass wearers. The use of a Barlow lens can improve the

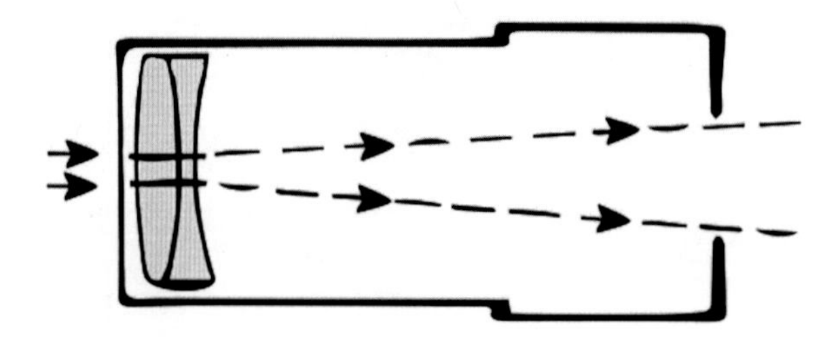

Fig. A15 The Barlow lens (Adam Chen)

off-axis edge sharpness of some eyepiece designs. The Barlow lens and related focal extenders are also useful for astrophotography.

Some discussion is needed on the subject of zoom eyepieces. In the 1960s, the zoom eyepiece earned a reputation for mediocre optics and was not worth the money. Today's zoom eyepieces deserve some attention. Improvements in lens coatings, the introduction of high index glass, and improved manufacturing has yielded a modern zoom that is worthy of a spot in an observer's eyepiece case. Although still narrower in field of view at longer focal lengths, and wider at shorter focal lengths, the performance has been greatly improved. A zoom eyepiece will not take the place of an eyepiece collection for critical observing, but serves the role for quick look situations, or when showing the night sky to children whose short attention spans don't allow for the changing and refocusing of conventional eyepieces to change magnification.

The driving criteria for eyepiece selection for lunar and planetary observing are sharpness and contrast. The rule of thumb for selecting the right eyepieces for viewing the planets and the Moon is "the simpler the better." The classic Abbe, Plossl, and Brandon designs are the preferred choices. There are more esoteric lunar and planetary eyepiece designs based on the monocentric design, or on proprietary designs. These are not discussed here due to their low availability.

There is an Achilles heel to the three classic orthoscopic designs. The older eyepiece designs perform best in longer focal length telescopes. In the era in which these designs originated, telescopes had long focal lengths, typically *f*/10 or greater. At *f*/20, even the lowly Ramsden design performs well. But many of today's telescopes have much shorter focal lengths, often *f*/6 or shorter. The classic designs suffer from loss of edge sharpness because of the steeper angle of the light cone from the objective as it enters the eyepiece. Modern designs take into account the shorter focal length telescopes of today. The design rationale for many of the updated configurations of the Plossl design has been to widen the field of view and improve performance with short focal length telescopes.

The recommendation for the ideal eyepiece for general observing is as follows:

1. For telescopes with a focal length of *f*/7 or greater, the Abbe or Plossl designs will perform at the highest level.
2. For telescopes with focal lengths of less than *f*/6, the modern wide-field and ultra-wide field designs are suggested.

3. For observers who must wear glasses while viewing, there are some proprietary eyepiece designs that provide 20 mm of eye relief. These tend to be premium eyepieces that use exotic lens configurations and glasses, and therefore are not cheap. But they are recommended for eyeglass wearers.
4. Consider using a Barlow in combination with a low- or medium-power eyepiece in order to obtain higher magnifications. The comfortable eye relief from this combination is often preferred by both eyeglass wearers and non-eyeglass wearers. The classic orthoscopic high-power (4 and 6 mm) eyepieces are notorious for their near-pinhole sized eye lens.

The Filters

There are a number of filters available to telescope owners, such as nebula filters, light pollution filters, and color filters. These filters are very useful in many applications where the goal is to reveal very dim low-contrast objects and features.

The question of light pollution filters and nebula filters is important, since the majority of the HST objects featured in this book are deep sky objects. These are remarkable telescope accessories that can produce a significant increase in contrast, and therefore observability, or dim deep sky objects. These filters are made of many layers of specialized coatings, with each coating formulated to filter a specific wavelength of light. It is important to remember that light pollution filters and nebula filters do not brighten the object, but darken the background and increase the contrast.

In general, light pollution reduction filters, often referred to as LPR filters, are broadband filters designed to filter the wavelengths of light associated with high- and low-pressure sodium streetlights, mercury vapor streetlights, and various wavelengths associated with houselights. These filters are no substitute for dark skies, but darken the background view through an eyepiece to the point of making a barely visible nebula or planetary readily visible. These are relatively mild filters when compared to the nebula filter cousins, and can be used with limited effectiveness in viewing galaxies and star clusters. Galaxies and star clusters are broadband objects, and the wavelengths of the streetlights will be filtered out of the objects frequency range, thus dimming the galaxy or cluster. The darker background may or may not be beneficial when viewing galaxies or clusters.

Nebula filters are designed and formulated to allow specific wavelengths of light through while suppressing most of the other light from an emission nebula or planetary nebula object. Nebula filters are useful for improving viewing of emission and planetary nebulas in dark skies as well as light polluted skies. Nebula filters take advantage of the physics of emission and planetary nebula. The nebula cloud atoms are ionized and excited into fluorescence causing the ghostly glow that characterizes these objects. Because the nebula cloud is composed of mostly ionized hydrogen atoms, the glow comes from the far-red and blue portion of the visible spectrum. Additionally, a characteristic green glow is emitted from two spikes of ionized oxygen. Therefore, narrowband nebula filters are formulated to allow oxygen-III, hydrogen-beta, or a combination of both to perform their magic. Commercially, these filters are known as O-III filter, H-beta filter, and the combination results in the term UHC filter, with UHC for Ultra High Contrast.

How well do these work? For example, the author used to live in an area with Washington, DC to the west, Baltimore to the north, and Annapolis MD to the east. To compound the light pollution problems caused by neighborhood streetlights, there was a shopping center ½ mile south and an elementary school to the north, both with mercury vapor parking lot lights. Needless to say, the sky was never really dark. In attempting to view the Owl Nebula in Ursa Major, even with an 11-in. SCT on a GoTo mount, the nebula could not be seen through the eyepiece without the aid of filtering. Adding a LPR filter darkened the ambient background to allow the Owl Nebula to be seen with averted vision (looking to the side of the object to take advantage of more sensitive areas of the eye). However, replacing the LPR with a UHC or O-III filter allowed the direct viewing of the Owl Nebula!

A word of caution to the astro-consumer: The hydrogen-beta filters are of limited use. In the northern hemisphere, the only objects for the backyard astronomer where the H-beta filter proves to be useful are the Horsehead Nebula and the California Nebula. There is more versatility to be gained in owning an LPR, UHC, and O-III filter.

Color filters can be used to enhance planetary details by suppressing a particular color and enhancing contrast. Appendix B details which Wratten-numbered filter to use for planetary details.

The Mounts

A solid telescope mount completes the total system needed for viewing the Moon, and beyond. There are two basic flavors of mounts: the altitude-azimuth mount, mostly referred to as the Alt-Azimuth or AltAz mount; and the equatorial mount. Each type can come either as manual, driven by hand controls or motors, and computer-driven GoTo models. Equatorial mounts and some computer-driven mounts compensate for the Earth's rotation and will track the Moon, planet, or other celestial object, thereby keeping the object in the field of view of the telescope (Fig. A16).

The most intuitive and easiest telescope mount is the alt-azimuth mount. Right-left and up and down. Simple in operation. In fact, it's the perfect mount for young people to use. Four-year-old kids have been seen at star parties using a refractor on an alt-az mount viewing the Moon with little supervision. Alt-azimuth mounts are considerably lighter than equatorial mounts, and are therefore well suited for grab-and-go scopes or for traveling. No setup is needed. The main drawback is the lack of tracking. The observer manually adjusts the positioning of the telescope, becoming the human tracking motors! (Figs. A17 and A18).

A notable example of an alt-azimuth mount is the implementation made famous by John Dobson in the early 1980s. Known as the Dobsonian mount (with the entire

Fig. A16 The author's telescopes featuring an Alt-Az with slow motion controls, two German equatorials, and computerized GoTo Mount (James Chen)

Fig. A17 Alt-Azimuth mount with slow motion controls (James Chen)

assembly including the Newtonian telescope being referred to as the Dobsonian telescope) is a simple, low center of gravity alt-azimuth mount made of wood and Teflon bearings. The Dobsonian caused a resurgence in homemade telescopes in the 1980s and 1990s. In today's market, telescope manufacturers dominate the

Fig. A18 The Dobsonian Mount (Hands-On-Optics archives)

12-in. Dobsonian and smaller sizes because of the economies of scale. Larger sizes are economically attractive for homebuilt projects, and for those with the funds, can be obtained as commercially produced telescopes. The Newtonian telescope on a Dobsonian mount offers by far the biggest "bang-for-the-buck." But they can be big and bulky, and the Newtonian optics still requires frequent alignment (Fig. A18).

With the exception of fork mounted Schmidt– and Maksutov–Cassegrains, the most popular form of equatorial mount in the amateur world is the German equatorial. The German mount is a tilted axis contraption with the right ascension axis pointed and aligned in the direction of the North Pole (for those down-under, the South Pole). A tracking motor applied to the right ascension axis drives the mount to keep the observed object in the eyepiece. German equatorial mounts are awkward and heavy. Care must be taken to balance the telescope on the mount, which explains the presence of the large counterweight that is a characteristic of the design. And polar aligning of this mount can be a chore. But if astrophotography is a goal, the German mount is a necessity (Fig. A19).

A GoTo telescope mount is quite simply a telescope system that is able to find celestial objects in the night sky, and then track them. The GoTo mount can be set

Fig. A19 German Equatorial Mount (James Chen)

Fig. A20 Computer GoTo Mounted Schmidt–Cassegrain (James Chen)

up in an alt-azimuth or equatorial fashion, and after the proper alignment procedure, the finderscope is not needed for the rest of the evening. Some of the newer GoTo telescopes have electronics that will perform the alignment procedure automatically (Fig. A20).

These telescope mounts are wonderful pieces of technology. The GoTo technology allows for more efficient use of observing time by quickly finding objects in the night sky. Built into the hand controller is a microprocessor, firmware, and built-in memory catalog of the positions of thousands of stars, galaxies, nebulae, open star clusters, globular clusters, planetary nebulae, our solar system planets, and of course the Moon. And the Moon is the one object that does not need a computer assist to find.

There is a trade-off when buying a GoTo mount. These mounts are not cheap. Often consumers are faced with the dilemma of either a smaller telescope with a GoTo mount, or a larger aperture telescope on a noncomputerized mount. In the case of viewing the Moon, a GoTo mount is not needed. If you can't find the Moon, it's either during the new moon phase, or you've got other problems! The Moon is an easy target.

A tip to GoTo owners: When using a telescope on an alt-azimuth GoTo mount, always use Polaris as one of your alignment stars. The computer algorithm used in programming the mount goes through less mathematical gymnastics when aligning with declination 0°, right ascension 0°. The GoTo accuracy is improved tenfold.

The Binoviewer Option

The majority of telescope owners make their observations through an eyepiece using one eye. The human brain is designed to process visual images through two eyes. There are two options for viewing the Moon, the planets and stars with two eyes. One is REALLY expensive—binocular telescopes. The other option is relatively affordable—the binoviewer. The binoviewer uses a system of prisms to split the single light path of a telescope into two separate light paths to two eyepieces (Fig. A21). This beam-splitting fools the eyes and the brain into thinking it is seeing an object in stereo. The results are spectacular when viewing the Moon. At certain high magnifications, and by allowing the Moon to drift through the field of view, the observer gets the sensation of orbiting the Moon and seeing the view that the Apollo command module pilot would see in orbit. With both eyes open, the lunar landscape seems to glide smoothly past. Even when tracking, the lunar landscape seems to take on three dimensions. The downside to owning a binoviewer is threefold:

1. There is a slight light loss using a binoviewer because of the additional light splitting optics. But for a bright object like the Moon, this is not a problem. For planetary views, the light loss is not of great impact. Deep sky observing can be problematic, especially with dim objects.
2. There is the additional expense of the binoviewer and buying two of every eyepiece. And you are limited to 1.25 in. sized eyepieces.
3. Many telescopes do not have enough in-focus to accommodate a binoviewer. SCTs and Maks focus by moving the primary mirror and binoviewers work well with these types. Some refractors are manufactured with shorter tubes to accommodate the binoviewer, and provide extension tubes to use for mono viewing.

Fig. A21 The binoviewer (James Chen)

Many binoviewers have an optional Barlow-like attachment to allow focusing with other types of telescopes, which limits the low power magnification range.

The Recommendation

Many of the readers of this book already own a telescope, on a stable mount, with a case or two of eyepieces and accessories.

To the readers without a telescope, get one. Astronomy is a wonderful hobby, filled with potential personal discoveries. For the price of a pair of bifocal high index eyeglasses, a nice 80 mm refractor on an alt-azimuth mount with two Plossl eyepieces can be obtained. This setup serves as a great care free introduction to the hobby, and when aperture fever takes hold (it always does) and a larger telescope is procured, the 80 mm refractor still has a role as a grab-and-go scope. The 80 mm refractor's sharp images are always appreciated.

However, the 80 mm refractor is light gathering challenged. There is a reason why large telescopes exist, they gather up much more light than small telescopes and make dim objects brighter. A telescope of 200 mm or greater is much more capable of observing the HST objects. A 200 or 250 mm SCT, Newtonian, or Dobsonian can produce very pleasing views of all the HST objects listed in this book.

At the risk of complicating matters, there is an argument to be made for the use of a smaller sized refractor instead of the larger but optically obstructed optical designs. Refractors by nature offer the high-contrast and sharpest images of all the optical designs. As a practical result, many objects seen in a 200 mm SCT can be equally observed visually with a 102–130 mm refractor, due to the tighter star images and higher contrast of the background.

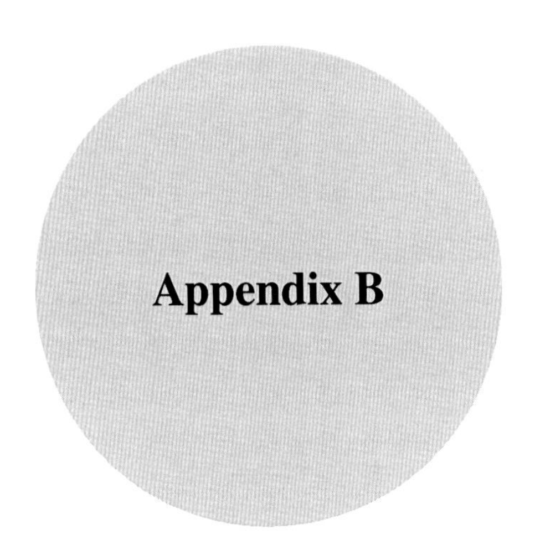

Color Filters Use

The numbering system for color filters is based on the photographic Wratten filters numbering system.

Color filters are used to enhance planetary details.

Darker filters are used for larger apertures, 200 mm or more.

Lighter filters are used with smaller apertures, 150 mm or less.

#8 Light Yellow
Moon: Feature Contrast
Mars: Maria
Jupiter: Belts
Jupiter: Orange-Red Zonal
Uranus: Dusky Detail
Neptune: Dusky Detail

#11 Yellow-Green
Mars: Maria
Jupiter: Clouds
Jupiter: Red/Blue Contrast
Saturn: Clouds
Saturn: Cassini Division
Saturn: Red/Blue Contrast

J.L. Chen, *Astronomy for Older Eyes*, The Patrick Moore Practical Astronomy Series, DOI 10.1007/978-3-319-52413-9

#12 Yellow
Moon: Feature Contrast
Mars: Blue-Green Areas
Jupiter: Red-Orange Features
Saturn: Clouds
Saturn: Red-Orange Features

#15 Dark Yellow
Moon: Feature Contrast
Mars: Clouds
Mars: Polar Caps
Jupiter: Belts
Saturn: Belts
Uranus: Dusky Detail
Neptune: Dusky

#21 Orange
Mars: Maria
Jupiter: Belts
Jupiter: Polar Regions
Saturn: Belts
Saturn: Polar Regions

#23A Light Red
Mercury: Planet/Sky Contrast
Mars: Maria
Mars: Blue-Green Areas
Jupiter: Belts
Jupiter: Polar Regions
Saturn: Belts
Saturn: Polar Regions

#25 Red
Mercury: Features
Venus: Planet/Sky Contrast
Venus: Terminator
Mars: Maria
Mars: Polar Caps
Jupiter: Belts
Jupiter: Galilean Moon Transits
Saturn: Clouds

#29 Dark Red
Mercury: Features
Venus: Planet/Sky Contrast
Venus: Terminator
Mars: Maria

Mars: Polar Caps
Jupiter: Belts
Jupiter: Galilean Moon Transits
Saturn: Clouds

#38A Dark Blue
Venus: Clouds
Mars: Dust Storms
Jupiter: Belts
Jupiter: Great Red Spot
Jupiter: Disc
Saturn: Belts

#47 Violet
Venus: Clouds
Mars: Polar Caps
Saturn: Rings

#80A Blue
Moon: Feature Contrast
Jupiter: Belts
Jupiter: Rilles
Jupiter: Festoons
Jupiter: Great Red Spot
Saturn: Belts
Saturn: Polar Regions

#82A Light Blue
Moon: Low-Contrast Features
Mars: Low-Contrast Features
Jupiter: Low-Contrast Features
Saturn: Low-Contrast Features

#56 Light Green
Moon: Detail
Mars: Dust Storms
Mars: Polar Caps
Jupiter: Belts
Jupiter: Atmosphere
Jupiter: Red/Blue/Light Contrast

#58 Green
Venus: Clouds
Mars: Polar Caps
Jupiter: Red/Blue/Light Contrast
Saturn: Belts
Saturn: Polar Regions

Common Telescope Formulas

Telescope Magnification

M = focal length of telescope/focal length of eyepiece

where the focal lengths of both telescope and eyepiece are in the same units.

M = f.l. telescope in mm/f.l. eyepiece in mm
or
M = f.l. telescope in inches/f.l. in inches

Exit Pupil

Exit Pupil = D/M

where

D = the diameter of the telescope's objective lens or primary mirror in millimeters
M = magnification = focal length of telescope/focal length of eyepiece
or Exit Pupil = F/f

where

F = the focal length of the eyepiece in millimeters
f = the telescope's focal ratio (the f-number)

J.L. Chen, *Astronomy for Older Eyes*, The Patrick Moore Practical Astronomy Series, DOI 10.1007/978-3-319-52413-9

Real Field of View

Real Field = F/M

where

F = the apparent field of view of the eyepiece
M = magnification

Focal Ratio

Focal Ratio = f.l./D

where

f.l. = focal length of the telescope
D = diameter of the telescope objective

Dawes Limit or Resolving Power

Estimate Resolving Power = 4.56/D in inches
or
Estimate Resolving Power = 116/D in mm

where

D = diameter of the telescope objective

Estimating Residual False Color in Achromatic Telescopes

From *Telescope Optics: Evaluation and Design,* by Harrie Rutten and Martin van Venrooij:

focal length $\geq$ 0.122D

where

D = diameter of the telescope objective in mm

If the focal length of an achromatic refractor is equal to or greater than this calculation, the residual chromatic aberration will not be a factor.

Astronomical League Observing Programs

Observing Programs Arranged by Equipment Needed to Complete Program. (source: astroleague.org)

Naked Eye Observing Programs (no equipment required)

Naked Eye Observing Programs
Analemma Observing Program
Comet Observing Program (possible to do, but will likely take many years)
Constellation Hunter Observing Programs
Dark Sky Advocate Observing Award
Earth Orbiting Satellite Observing Program
Meteor Observing Program
Outreach Observing Award

Binocular Observing Programs (no telescopes required)

Binocular Observing Programs
Advanced Binocular Double Star Program
Binocular Double Star Observing Program
Binocular Messier Observing Program
Binocular Variable Star Program
Comet Observing Program
Deep Sky Binocular Observing Program
Outreach Observing Award
Southern Skies Binocular Program

J.L. Chen, *Astronomy for Older Eyes*, The Patrick Moore Practical Astronomy Series, DOI 10.1007/978-3-319-52413-9

Telescopic Observing Programs

Active Galactic Nuclei Observing Program
Arp Peculiar Galaxies Northern Observing Program
Arp Peculiar Galaxies Southern Observing Program
Asterism Observing Program
Asteroid Observing Program
Bright Nebula Observing Program
Caldwell Observing Program
Carbon Star Observing Program
Comet Observing Program
Dark Nebulae Observing Program
Double Star Observing Program
Flat Galaxies Observing Program
Galaxy Group and Clusters Observing Program
Galileo Observing Program
Globular Cluster Observing Program
Herschel 400 Observing Program
Herschel II Observing Program
Hydrogen Alpha Solar Observing Program (a hydrogen-alpha telescope is required)
Local Galaxy Groups and Neighborhood Observing Program
Lunar Observing Program
Lunar II Observing Program
Master Observer Observing Award
Messier Observing Program
NEO Observing Award
NASA Observing Challenges Special Observing Awards
Occultation Observing Program
Open Cluster Observing Program
Outreach Observing Award
Planetary Nebula Observing Program
Planetary Transit Special Observing Awards (Venus and Mercury)
Radio Astronomy Observing Program (uses radio telescopes)
Sketching Observing Award
Sky Puppy Observing Program
Solar System Observing Program
Southern Sky Telescopic Observing Program
Sunspotters Observing Program (a solar filter is required)
Stellar Evolution Observing Program
Two in the View Observing Program
Universe Sampler observing Program
Urban Observing Program
Variable Star Observing Program

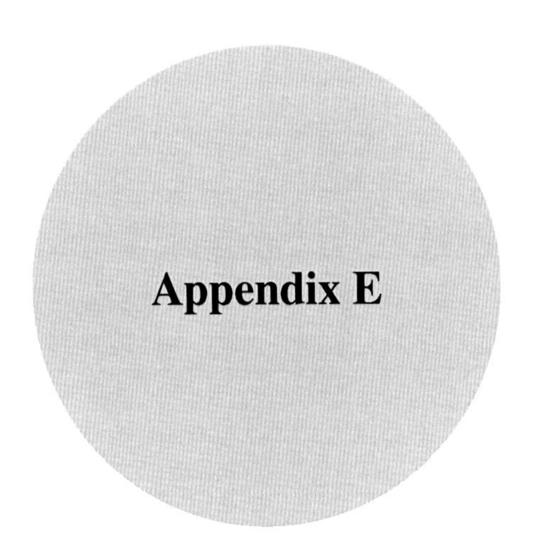

North America Star Parties

This is a list of major star parties in North America, as compiled by *Sky and Telescope* magazine, and provided here.

Be aware that there are a number of small regional star parties and public outreach programs across the country that are supported by local astronomy clubs that are not listed here.

Please note: The Northeast Astronomy Forum is listed as a star party. NEAF, as it is called, is not a traditional star party where amateur astronomers gather to observe the nighttime sky. NEAF is known for its science briefings provided by nationally and world-renown professional astronomers, scientists, and engineers. NEAF is also known for its 2-day telescope expo with telescope vendors and telescope shops on hand to display and sell their products, often at a good discount.

Name	Location	Date
Winter Star Party	West Summerland Key, FL	Feb
Hodges Gardens Star Party	Florien, LA	Mar
Delmarva Spring Star Gaze	Tuckahoe State Park, MD	Mar
Southern Star Astronomical Convention	Little Switzerland, NC	Apr
Mid-South Star Gaze	French Camp, MS	Apr
Northeast Astronomy Forum	Suffern, NY	Apr
Astronomy Day (spring)	Everywhere!	Apr/May

(continued)

J.L. Chen, *Astronomy for Older Eyes*, The Patrick Moore Practical Astronomy Series, DOI 10.1007/978-3-319-52413-9

(continued)

Name	Location	Date
Texas Star Party	Fort Davis, TX	May
TSSP-TN Spring Star Party	Fall Creek Falls State Park, TN	May
RTMC Astronomy Expo	Riverside, PA	Memorial Day weekend
Cherry Springs Star Party	Coudersport, PA	June
Grand Canyon Star Party	Grand Canyon, AZ	June
Rocky Mountain Star Stare	Gardner, CO	June
Gateway to the Universe Star Party	Marten River Provincial Park, ON	July
Nebraska Star Party	Valentine, NE	July
Golden State Star Party	Bieber, CA	July
Mason-Dixon Star Party	York, PA	July
Mount Kobau Star Party	Osoyoos, BC	Aug
Table Mountain Star Party	Oroville, WA	Aug
Oregon Star Party	Indian Trail Spring, OR	Aug
Stellafane Convention	Springfield, VT	Aug
StarFest	Ayton, ON	Aug
Nova East	Smileys Provincial Park, NS	Aug
Manitoulin Star Party	Manitoulin Island, ON	Aug
Almost Heaven Star Party	Spruce Knob, WV	Aug
Black Forest Star Party	Coudersport, PA	Sep
AstroAssembly	North Scituate, RI	Sep
Idaho Star Party	Bruneau, ID	Sep
Alberta Star Party	Caroline, AB	Sep
Okie-Tex Star Party	Kenton, OK	Sep
Enchanted Skies Star Party	Socorro, NM	Oct
Astronomy Day (autumn)	Everywhere!	Sep/Oct
Twin Lakes Star Party	KY	Oct
SJAC Fall Star Party	Belleplain State Forest, NJ	Oct
Deep South Regional Star Gaze	McComb, MS	Oct/Nov
Mid-Atlantic Star Party	Robbins, NC	Oct/Nov
Chiefland Star Party	Chiefland Astronomy Village, FL	Nov

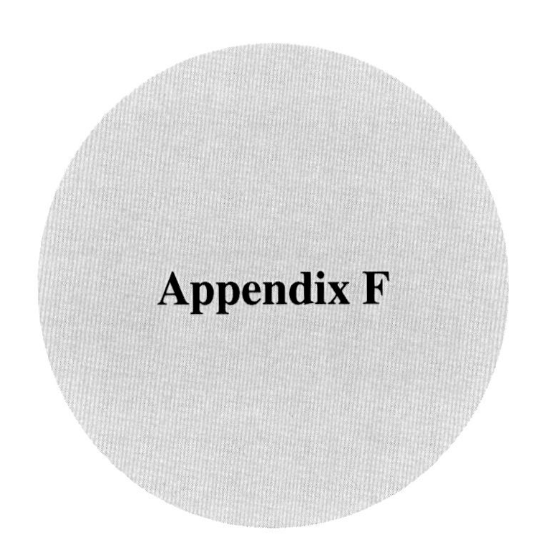

Appendix F

Messier Catalog

During his lifetime, Charles Messier (1730–1817) was an astronomer noted for his comet discoveries. He found 13 comets and shared in seven more co-discoveries. Messier compiled a list of deep sky objects that were easily confused for comets to help him in his comet searches. Ironically, he is more famous today for his list of non-comet deep sky objects than his comet discoveries. Known as the Messier catalog, it contains 110 objects (actually 109 because of a duplication), including nebulae, clusters, and galaxies. All of these objects can be seen through an 80 mm refractor or a pair of 50 mm binoculars.

Messier number	Common name	Constellation	R.A. H:M.S	DEC	App Mag	Type
M1	Crab Nebula	Taurus	5:34.5	22°01′	8.4	Planetary Nebula
M2		Aquarius	21:33.5	−00°49′	6.5	Globular Cluster
M3		Canes Venatici	13:42.2	28°23′	6.4	Globular Cluster
M4		Scorpius	16:23.6	−26°32′	5.9	Globular Cluster
M5		Serpens	15:18.5	2°05′	5.8	Globular Cluster

(continued)

J.L. Chen, *Astronomy for Older Eyes*, The Patrick Moore Practical Astronomy Series, DOI 10.1007/978-3-319-52413-9

(continued)

Messier number	Common name	Constellation	R.A. H:M.S	DEC	App Mag	Type
M6	Butterfly Cluster	Scorpius	17:40.0	−32°13′	4.2	Open Cluster
M7	Ptolemy Cluster	Scorpius	17:54.0	−34°49′	3.3	Open Cluster
M8	Lagoon Nebula	Sagittarius	18:03.7	−24°23′	5.8	Emission Nebula
M9		Ophiuchus	17:19.2	−18°31′	7.9	Globular Cluster
M10		Ophiuchus	16:57.2	−4°06′	6.6	Globular Cluster
M11	Wild Duck Cluster	Scutum	18:51.1	−6°16′	5.8	Open Cluster
M12		Ophiuchus	16:47.2	−1°57′	6.6	Globular Cluster
M13	Hercules Cluster	Hercules	16:41.7	36°28′	5.9	Globular Cluster
M14		Ophiuchus	17:37.6	−3°15′	7.6	Globular Cluster
M15		Pegasus	21:30.0	12°10′	6.4	Globular Cluster
M16	Eagle Nebula	Serpens	18:18.9	−13°47′	6	Emission Nebula
M17	Omega, Swan, Horseshoe, or Lobster Nebula	Sagittarius	18:20.8	−16°11′	7	Nebula
M18		Sagittarius	18:19.9	−17°08′	6.9	Open Cluster
M19		Ophiuchus	17:02.6	−26°16′	7.2	Globular Cluster
M20	Trifid Nebula	Sagittarius	18:02.4	−23°02′	8.5	Diffuse Nebula
M21		Sagittarius	18:04.7	−22°30′	5.9	Open Cluster
M22	Sagittarius Cluster	Sagittarius	18:36.4	−23°54′	5.1	Globular Cluster
M23		Sagittarius	17:56.9	−19°01′	5.5	Open Cluster
M24	Sagittarius Star Cloud	Sagittarius	18:16.4	−18°29′	4.5	Open Cluster
M25		Sagittarius	18:31.7	−19°15′	4.6	Open Cluster
M26		Scutum	18:45.2	−9°24′	8	Open Cluster
M27	Dumbbell Nebula	Vulpecula	19:59.6	22°43′	8.1	Planetary Nebula
M28		Sagittarius	18:24.6	−24°52′	6.9	Globular Cluster
M29		Cygnus	20:23.0	38°32′	6.6	Open Cluster
M30		Capricornus	21:40.4	−23°11′	7.5	Globular Cluster
M31	Andromeda Galaxy	Andromeda	0:42.7	41°16′	3.4	Spiral Galaxy
M32		Andromeda	0:42.7	40°52′	8.2	Elliptical Galaxy

(continued)

(continued)

Messier number	Common name	Constellation	R.A. H:M.S	DEC	App Mag	Type
M33	Pinwheel Galaxy	Triangulum	1:33.8	30°39′	5.7	Spiral Galaxy
M34		Perseus	2:42.0	42°47′	5.2	Open Cluster
M35		Gemini	6:08.8	24°20′	5.1	Open Cluster
M36		Auriga	5:36.3	34°08′	6	Open Cluster
M37		Auriga	5:52.0	32°33′	5.6	Open Cluster
M38		Auriga	5:28.7	35°50′	6.4	Open Cluster
M39		Cygnus	21:32.3	48°26′	4.6	Open Cluster
M40	Winnecke 4	Ursa Major	12:22.2	68°05′	8	Dbl star
M41		Canis Major	6:47.0	−20°44′	4.5	Open Cluster
M42	Great Orion Nebula	Orion	5:35.3	−5°27′	4	Nebula
M43	De Mairan's Nebula	Orion	5:35.5	−5°16′	9	Nebula
M44	Beehive Cluster	Cancer	8:40.0	19°59′	3.1	Open Cluster
M45	Pleiades	Taurus	3:47.5	24°07′	1.2	Open Cluster
M46		Puppis	7:41.8	−14°49′	6.1	Open Cluster
M47		Puppis	7:36.6	−14°30′	4.4	Open Cluster
M48		Hydra	8:13.8	−5°48′	5.8	Open Cluster
M49		Virgo	12:29.8	8°00′	8.4	Elliptical galaxy
M50		Monoceros	7:03.0	−8°20′	5.9	Open Cluster
M51	Whirlpool Galaxy	Canes Venatici	13:29.9	47°12′	8.1	Spiral Galaxy
M52		Cassiopeia	23:24.2	61°35′	6.9	Open Cluster.
M53		Coma Berenices	13:12.9	18°10′	7.7	Globular Cluster
M54		Sagittarius	18:55.1	−30°29′	7.7	Globular Cluster
M55		Sagittarius	19:40 .0	−30°68′	7	Globular Cluster
M56		Lyra	19:16.6	30°11′	8.2	Globular Cluster
M57	Ring Nebula	Lyra	18:53.6	33°02′	9	Planetary Nebula
M58		Virgo	12:37.7	11°49′	9.8	Spiral Galaxy
M59		Virgo	12:42.0	11°39′	9.8	Elliptical Galaxy
M60		Virgo	12:43.7	11°33′	8.8	Elliptical Galaxy
M61		Virgo	12:21.9	4°28′	9.7	Spiral Galaxy
M62		Ophiuchus	17:01.2	−30°07′	6.6	Globular Cluster

(continued)

(continued)

Messier number	Common name	Constellation	R.A. H:M.S	DEC	App Mag	Type
M63	Sunflower Galaxy	Canes Venatici	13:15.8	42°02′	8.6	Spiral Galaxy
M64	Black Eye Galaxy	Coma Berenices	12:56.7	21°41′	8.5	Spiral Galaxy
M65	Leo's Triplet	Leo	11:18.9	13°05′	9.3	Spiral Galaxy
M66	Leo's Triplet	Leo	11:20.3	12°59′	9	Spiral Galaxy
M67		Cancer	8:50.3	11°49′	6.9	Open Cluster
M68		Hydra	12:39.5	−26°45′	8.2	Globular Cluster
M69		Sagittarius	18:31.4	−32°21′	7.7	Globular Cluster
M70		Sagittarius	18:43.2	−32°18′	8.1	Globular Cluster
M71		Sagitta	19:53.7	18°47′	8.3	Globular Cluster
M72		Aquarius	20:53.5	−12°32′	9.4	Globular Cluster
M73		Aquarius	20:68.0	−12°38′		asterism
M74		Pisces	1:36.7	15°47′	9.2	Spiral Galaxy
M75		Sagittarius	20:06.1	−21°55′	8.6	Globular Cluster
M76	Cork Nebula, Little Dumbbell	Perseus	1:42.2	51°34′	11.5	Planetary Nebula
M77		Cetus	2:42.7	0°01′	8.8	Spiral Galaxy
M78		Orion	5:46.7	0°03′	8	Nebula
M79		Lepus	5:24.2	−24°33′	8	Globular Cluster
M80		Scorpius	16:17.0	−22°59′	7.2	Globular Cluster
M81	Bodes Nebula	Ursa Major	9:55.8	69°04′	6.8	Spiral Galaxy
M82	Cigar Galaxy	Ursa Major	9:56.2	69°41′	8.4	Irregular Galaxy.
M83	Southern Pinwheel Galaxy	Hydra	13:37.7	−29°52′	7.6	Spiral Galaxy
M84		Virgo	12:25.1	12°53′	9.3	Elliptical Galaxy
M85		Coma Berenices	12:25.4	18°11′	9.2	Elliptical Galaxy
M86		Virgo	12:26.2	12°57′	9.2	Elliptical Galaxy

(continued)

(continued)

Messier number	Common name	Constellation	R.A. H:M.S	DEC	App Mag	Type
M87	Virgo A	Virgo	12:30.8	12°24′	8.6	Elliptical Galaxy
M88		Coma Berenices	12:32.0	14°25′	9.5	Spiral Galaxy
M89		Virgo	12:35.7	12°33′	9.8	Elliptical Galaxy
M90		Virgo	12:36.8	13°10′	9.5	Spiral Galaxy
M91		Coma Berenices	12:35.4	14°30′	10.2	Spiral Galaxy
M92		Hercules	17:17.1	43°08′	6.5	Globular Cluster
M93		Puppis	7:44.6	−23°52′	6.2	Open Cluster
M94		Canes Venatici	12:50.9	41°07′	8.1	Spiral Galaxy
M95		Leo	10:44.0	11°42′	9.7	Spiral Galaxy
M96		Leo	10:46.8	11°49′	9.2	Spiral Galaxy
M97	Owl Nebula	Ursa Major	11:14.9	55°01′	11	Planetary Nebula
M98		Coma Berenices	12:13.8	14°54′	10.1	Spiral Galaxy
M99	Pin Wheel Nebula	Coma Berenices	12:18.8	14°25′	9.8	Spiral Galaxy
M100		Coma Berenices	12:22.9	15°49′	9.4	Spiral Galaxy
M101		Ursa Major	14:03.2	54°21′	7.7	Spiral Galaxy
M102	Probably M101 duplicate		14:03.2	54°21′	7.7	Duplicate
M103		Cassiopeia	1:33.1	70°42′	7.4	Open Cluster
M104	Sombrero Galaxy	Virgo	12:40.0	−11°37′	8.3	Spiral Galaxy
M105		Leo	10:47.9	12°35′	9.3	Elliptical Galaxy
M106		Canes Venatici	12:19.0	47°18′	8.3	Spiral Galaxy
M107		Ophiuchus	16:32.5	−13°03′	8.1	Globular Cluster
M108		Ursa Major	11:11.6	55°40′	10	Spiral Galaxy
M109		Ursa Major	11:57.7	53°23′	9.8	Spiral Galaxy
M110		Andromeda	0:40.3	41°41′	8	Elliptical Galaxy

Selected Non-Messier Catalog NGC Objects

Amateur astronomers often challenge themselves by tackling J. L. E. Dreyer's New General Catalogue of Nebulas and Cluster of Stars (NGC) and the two supplements, the Index Catalogues (IC) in their databases.

The NGC list, compiled in 1888, contains 7840 objects. The two supplements were published in 1895 and 1908, respectively, with the first containing 1520 additional deep sky objects and the second containing 3866 additional IC objects.

Many of the objects listed in the NGC/IC database of the both Star Books are not visible visually through the size of telescope that can be accommodated by the Sphinx family of mounts. It is possible to detect and image the fainter objects photographically.

Unlike the much shorter Messier Catalog of Appendix B, it is impractical to list the thousands of NGC/IC objects in this book.

However, here are some NGC and IC objects that are recommended. This list is called the SAA 100.

A discussion thread began in 2000 in the sci.astro.amateur (SAA) newsgroup when a question was posted: "What are your favorite non-Messier objects for 8–12" telescopes?" The SAA newsgroup participants responded enthusiastically to the question, posting many messages nominating a wide variety of objects.

The table below lists the SAA 100 in rank order by number of votes received. About half the objects received only one vote each; these are listed alphabetically at the end of the list.

J.L. Chen, *Astronomy for Older Eyes*, The Patrick Moore Practical Astronomy Series, DOI 10.1007/978-3-319-52413-9

This is by no means the only available list of non-Messier objects, with the Caldwell List coming quickly to mind. The SAA 100 is a good starting point for readers of this book.

SAA 100

Best Non-Messier Objects, in Rank Order Object types: Gal = galaxy, OC = open cluster, GC = globular cluster, PN = planetary nebula, BN = bright nebula, DN = dark nebula, SNR = supernova remnant

Object	Type	Con	VisualMag	Size	RA	Dec	Pop. name	Notes
NGC 253	Gal	Scl	7.2	25.0′ × 7.0′	00h 47m 35s	−25° 17′ 01″		Edge-on spiral
NGC 4565	Gal	Com	9.6	15.5′ × 1.9′	12h 36m 20s	+25° 59′ 23″	Bernice's Hair Clip	Classic edge-on spiral with dust lane
NGC 6960	SNR	Cyg		70.0′ × 6.0′	20h 45m 38s	+30° 43′ 20″	Western Veil	
NGC 6992	SNR	Cyg		25.0′ × 20.0′	20h 56m 14s	+31° 04′ 20″	Eastern Veil	
NGC 869	OC	Per	5.3	30.0′	02h 19m 03s	+57° 08′ 58″	Double Cluster	w/NGC 884
NGC 884	OC	Per	6.1	30.0′	02h 22m 27s	+57° 06′ 57″	Double Cluster	w/NGC 869
NGC 457	OC	Cas	6.4	13.0′	01h 19m 10s	+58° 20′ 02″	Owl Cluster	
NGC 5139	GC	Cen	3.7	36.3′	13h 26m 46s	−47° 28′ 45″	Omega Centauri	Best GC in the sky
NGC 7293	PN	Aqr	6.3	16.0′ × 12.0′	22h 29m 40s	−20° 47′ 23″	Helical Nebula	
NGC 7789	OC	Cas	6.7	16.0′	23h 57m 04s	+56° 44′ 09″		
NGC 2237	BN	Mon	5.5	70.0′ × 80.0′	06h 32m 19s	+04° 59′ 03″	Rosette Nebula	OC NGC 2244 embedded in nebula
NGC 2244	OC	Mon	4.8	24.0′	06h 32m 25s	+04° 52′ 03″		Involved with Rosette Neb. (NGC 2237)
NGC 2359	BN	CMa		8.0′	07h 17m 48s	−13° 12′ 54″	Thor's Helmet; Duck Nebula	Wolf–Rayet remnant
NGC 2392	PN	Gem	8.6	47.0″ × 43.0″	07h 29m 10s	+20° 54′ 42″	Eskimo Nebula; Clown Face	
NGC 3242	PN	Hya	8.6	40.0″ × 35.0″	10h 24m 48s	−18° 38′ 14″	Ghost of Jupiter	
NGC 6543	PN	Dra	8.3	22.0″ × 16.0″	17h 58m 36s	+66° 38′ 17″	Cat's Eye Nebula	
NGC 4631	Gal	CVn	9.2	17.0′ × 3.5′	12h 42m 11s	+32° 32′ 42″		Same LP field as NGC 4656
NGC 4656	Gal	CVn	10.5	22.0′ × 3.0′	12h 43m 58s	+32° 10′ 21″		Same LP field as NGC 4631
NGC 5128	Gal	Cen	6.8	18.2′ × 14.5′	13h 25m 29s	−43° 01′ 07″	Centaurus A	Strong radio source
NGC 6781	PN	Aql	11.8	1.9′ × 1.8′	19h 18m 28s	+06° 32′ 46″		
NGC 6826	PN	Cyg	8.8	27.0″ × 24.0″	19h 44m 53s	+50° 31′ 42″	Blinking Planetary	
NGC 7009	PN	Aqr	8.3	28.0″ × 23.0″	21h 04m 15s	−11° 21′ 49″	Saturn Nebula	

(continued)

(continued)

Object	Type	Con	VisualMag	Size	RA	Dec	Pop. name	Notes
Abell 1656	Gal cluster	Com	11.0	120.0′	12h 59m 48s	+27° 59′ 04″	Coma Gal Cluster	Tough in 8–12″ aperture
NGC 1023	Gal	Per	9.4	9.0′ × 4.0′	02h 40m 27s	+39° 03′ 47″		
NGC 2362	OC	CMa	4.1	8.0′	07h 18m 48s	−24° 56′ 51″		
NGC 2403	Gal	Cam	8.5	17.8′	07h 36m 55s	+65° 35′ 42″		
NGC 4038	Gal	Cor	10.3	2.6′ × 1.8′	12h 01m 53s	−18° 51′ 55″	The Antennae; Ringtail Gal	Interacting with NGC 4039
NGC 4039	Gal	Cor	10.6	3.2′ × 2.2′	12h 01m 54s	−18° 53′ 07″	The Antennae	Interacting with NGC 4038
NGC 5907	Gal	Dra	10.3	12.8′ × 1.8′	15h 15m 52s	+56° 19′ 48″		
NGC 6369	PN	Oph	11.0	30.0″ × 29.0″	17h 29m 22s	−23° 45′ 37″		
NGC 663	OC	Cas	7.1	16.0′	01h 46m 04s	+61° 15′ 00″		
NGC 654	OC	Cas	6.5	5.0′	01h 44m 10s	+61° 53′ 00″		
NGC 659	OC	Cas	7.9	5.0′	01h 44m 16s	+60° 42′ 00″		
NGC 7000	BN	Cyg		175.0′ × 110.0′	20h 58m 32s	+44° 33′ 21″	North American Nebula	Large; often easier in binoculars than telescope
NGC 7331	Gal	Peg	9.5	11.4′ × 4.0′	22h 37m 08s	+34° 25′ 27″	Little And Gal	
NGC 7662	PN	And	8.6	17.0″ × 14.0″	23h 25m 57s	+42° 32′ 44″	Blue Snowball Nebula	
B 59, 65-7	DN	Oph		300.0′	17h 21m 02s	−26° 59′ 58″	Pipe Nebula (stem)	
B 78	DN	Oph		200.0′	17h 33m 02s	−25° 59′ 58″	Pipe Nebula (bowl)	
IC 1396	BN	Cep	3.5	154.0′ × 140.0′	21h 39m 09s	+57° 46′ 58″		
IC 418	PN	Lep	10.7	14.0″ × 11.0″	05h 27m 30s	−12° 41′ 32″		
IC 4665	OC	Oph	4.2	41.0′	17h 46m 20s	+05° 43′ 08″		
Mel 111	OC	Com	1.8	275.0′	12h 25m 00s	+26° 00′ 07″	Coma Berenices Star Cluster	
Mel 20	OC	Per	1.2	185.0′	03h 22m 03s	+48° 59′ 56″	Alpha Persei Association	

NGC 1502	OC	Cam	6.9	8.0′	04h 07m 45s	+62° 19′ 49″		Near SE end of Kemble's Cascade
NGC 1528	OC	Per	6.4	24.0′	04h 15m 24s	+51° 13′ 49″		
NGC 1907	OC	Aur	8.2	7.0′	05h 28m 00s	+35° 18′ 53″		
NGC 1973	BN	Ori		5.0′ × 5.0′	05h 35m 09s	−04° 43′ 56″	Part of Running Man Nebula	
NGC 1975	BN	Ori		10.0′ × 5.0′	05h 35m 21s	−04° 40′ 56″	Part of Running Man Nebula	
NGC 1977	BN	Ori		20.0′ × 10.0′	05h 35m 27s	−04° 49′ 56″	Part of Running Man Nebula	42 Orionis nebula
NGC 2070	BN	Dor	8.3	5.0′	05h 38m 39s	−69° 04′ 51″	Tarantula Nebula	In Lg. Magellanic Cloud
3C 273	Quasar	Vir	12.0		12h 29m 06s	+02° 03′ 01″		Brightest quasar; most remote object visible in modest amateur telescopes (~ two billion light years)
Albireo	Star	Cyg	3.1		19h 30m 45s	+27° 57′ 55″		Superb double star; blue-white/yellow
Cr 399	Asterism	Vul	3.6	60.0′	19h 25m 26s	+20° 11′ 18″	Brocchi's Cluster; The Coathanger	Once assumed to be a OC; data from Hipparcos spacecraft shows it to be a chance alignment of stars
Fornax Gal. Cluster	Gal cluster	For		3° × 2°	03h 38m 31s	−35° 26′ 40″		Approx. 20 galaxies brighter than mag. 13
Kemble's Cascade	Asterism	Cam			03h 57m 30s	+63° 04′ 13″		First described by Canadian amateur Lucian J. Kemble; beautiful chain of about 20 mag. 5…9 stars; coordinates are for SAO 12969, a mag. 5 star in the middle of the Cascade
King 10	OC	Cep		3.0′	22h 54m 58s	+59° 10′ 16″		

(continued)

(continued)

Object	Type	Con	VisualMag	Size	RA	Dec	Pop. name	Notes
Markarian's Chain	Gal chain	Vir			12h 25m 04s	+12° 53′ 16″		String of bright galaxies; covers 3° of sky, starting with M84 and M86 in Virgo, ending with NGCs 4459 and 4474 in Coma Berenices; coordinates are for M 84
Mel 25	OC	Tau	0.5	330.0′	04h 27m 02s	+16° 00′ 03″	Hyades	Aldebaran not a member
NGC 104	GC	Tuc	4.0	30.9′	00h 24m 10s	−72° 04′ 37″	47 Tucanae	
NGC 1535	PN	Eri	10.4	20.0″ × 17.0″	04h 14m 16s	−12° 44′ 16″		Multiple shells
NGC 2158	OC	Gem	8.6	5.0′	06h 07m 33s	+24° 05′ 56″		
NGC 2169	OC	Ori	5.9	7.0′	06h 08m 27s	+13° 56′ 59″	"37" Cluster	
NGC 2174	BN	Ori		25.0′ × 20.0′	06h 10m 01s	+20° 33′ 58″		
NGC 2232	OC	Mon	3.9	30.0′	06h 26m 37s	−04° 44′ 54″		
NGC 225	OC	Cas	7.0	12.0′	00h 43m 28s	+61° 47′ 06″		
NGC 2261	BN	Mon		2.0′ × 1.0′	06h 39m 13s	+08° 44′ 01″	Hubble's Variable Nebula	
NGC 2264	OC	Mon	3.9	30.0′ × 60.0′	06h 40m 58s	+09° 53′ 42″	Christmas Tree Cluster; Cone Nebula	Includes naked-eye S Mon (15 Mon)
NGC 2301	OC	Mon	6.0	12.0′	06h 51m 49s	+00° 28′ 04″		
NGC 2360	OC	CMa	7.2	13.0′	07h 17m 48s	−15° 36′ 53″		
NGC 2438	PN	Pup	11.0	1.1′	07h 41m 51s	−14° 44′ 06″		In foreground of M 46
NGC 2467	BN	Pup	7.1	15.0′	07h 52m 30s	−26° 22′ 52″		Use UHC or O-III filter; includes loose cluster of mag. 8–12 stars
NGC 247	Gal	Cet	9.1	20.0′ × 7.0′	00h 47m 11s	−20° 45′ 21″		
NGC 2841	Gal	Uma	9.2	7.4′ × 3.5′	09h 22m 01s	+50° 58′ 21″		
NGC 2903	Gal	Leo	9.0	13.3′ × 6.0′	09h 32m 10s	+21° 29′ 58″		
NGC 3115	Gal	Sex	8.9	8.3′ × 3.2′	10h 05m 14s	−07° 43′ 06″	Spindle Gal	

NGC 3372	BN	Car		120.0′ × 120.0′	10h 43m 47s	−59° 52′ 01″	Eta Carina Nebula	
NGC 3532	OC	Car	3.0	55.0′	11h 06m 23s	−58° 40′ 03″		
NGC 3766	OC	Cen	5.3	12.0′	11h 36m 05s	−61° 37′ 04″		
NGC 3877	Gal	Uma	11.0	5.6′ × 1.2′	11h 46m 07s	+47° 29′ 37″		
NGC 40	PN	Cep	10.7	1.0′ × 0.7′	00h 13m 08s	+72° 31′ 47″		
NGC 4244	Gal	CVn	10.4	18.5′ × 2.3′	12h 17m 29s	+37° 48′ 28″		
NGC 4361	PN	Cor	10.3	1.3′	12h 24m 30s	−18° 47′ 38″		
NGC 4526	Gal	Vir	9.7	7.0′ × 2.7′	12h 34m 03s	+07° 42′ 03″	Lost Gal	
NGC 4567	Gal	Vir	11.3	3.0′ × 2.5′	12h 36m 33s	+11° 15′ 33″	Siamese Twins	Overlaps NGC 4568
NGC 4568	Gal	Vir	10.8	5.1′ × 2.4′	12h 36m 35s	+11° 14′ 17″	Siamese Twins	Overlaps NGC 4567
NGC 4755	OC	Cru	4.2	10.0′	12h 53m 35s	−60° 20′ 08″	Jewel Box Cluster; Kappa Crucis	
NGC 5746	Gal	Vir	10.3	7.4′ × 1.1′	14h 44m 57s	+01° 57′ 20″		
NGC 6210	PN	Her	9.7	20.0″ × 13.0″	16h 44m 30s	+23° 48′ 46″		
NGC 6231	OC	Sco	2.6	15.0′	16h 54m 01s	−41° 48′ 06″	Table of Scorpius	Zeta Sco complex
NGC 6397	GC	Ara	5.7	25.7′	17h 40m 43s	−53° 40′ 33″		One of the nearest globulars
NGC 6545	Gal	Pav	13.2	1.0′ × 0.9′	18h 12m 18s	−63° 46′ 45″	Needle Galaxy	
NGC 6572	PN	Oph	9.0	15.0″ × 12.0″	18h 12m 09s	+06° 51′ 01″		
NGC 6633	OC	Oph	4.6	27.0′	18h 27m 43s	+06° 34′ 14″		
NGC 6819	OC	Cyg	7.3	5.0′	19h 41m 20s	+40° 11′ 22″		
NGC 6885	OC	Vul	8.1	7.0′	20h 12m 02s	+26° 29′ 20″		
NGC 6888	BN	Cyg		20.0′ × 10.0′	20h 12m 14s	+38° 20′ 21″	Crescent Nebula	
NGC 6939	OC	Cep	7.8	8.0′	20h 31m 27s	+60° 38′ 22″		
NGC 752	OC	And	5.7	50.0′	01h 57m 51s	+37° 41′ 05″		
NGC 891	Gal	And	9.9	14.0′ × 3.0′	02h 22m 36s	+42° 20′ 50″		Edge-on spiral w/prominent dust lane
Stock 2	OC	Cas	4.4	60.0′	02h 15m 04s	+59° 15′ 58″	Muscleman Cluster	

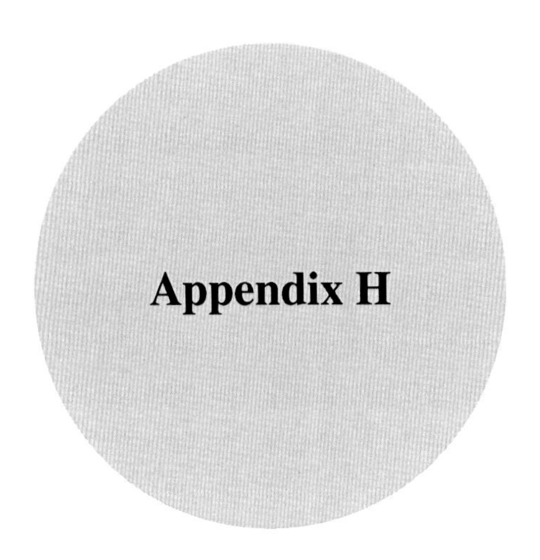

Appendix H

The Caldwell Catalog

Sir Patrick Caldwell-Moore, in 1995, noted that the Messier catalog does not include a number of bright deep sky objects, nor does it cover any Southern Hemisphere objects south of declination −35°. He compiled a new catalog to complement the famous Messier Catalog by including the "missing" objects and to extend the list to cover the Southern Hemisphere.

The resulting list became known as the Caldwell Catalog, which is a collection of 109 of the most impressive celestial objects culled from the NGC and IC catalogs that were not included in Messier's list. Objects in the Caldwell Catalog are organized in descending declination, while objects in Messier's Catalog are listed in order of discovery.

C#	NGC/IC	Con.	Type	R. A. h m	Dec. °…′	Mag.	Size ()	Description
1	188	Cep	OC	00 44.4	+85 20	8.1	14	
2	40	Cep	PN	00 13.0	+72 32	11.6	0.6	
3	4236	Dra	SbG	12 16.7	+69 28	9.7	21 × 7	
4	7023	Cep	BN	21 01.8	+68 12	6.8	18 × 18	Reflection Nebula
5	IC 342	Cam	SBcG	03 46.8	+68 06	9.2	18 × 17	
6	6543	Dra	PN	17 58.6	+66 38	8.8	0.3/5.8	Cat's Eye Nebula
7	2403	Cam	ScG	07 36.9	+65 36	8.9	18 × 10	
8	559	Cas	OC	01 29.5	+63 18	9.5	4	

(continued)

J.L. Chen, *Astronomy for Older Eyes*, The Patrick Moore Practical Astronomy Series, DOI 10.1007/978-3-319-52413-9

(continued)

C#	NGC/IC	Con.	Type	R. A. h m	Dec. °…′	Mag.	Size ()	Description
9	Sh2-155	Cep	BN	22 56.8	+62 37	7.7	50 × 10	Cave Nebula
10	663	Cas	OC	01 46.0	+61 15	7.1	16	
11	7635	Cas	BN	23 20.7	+61 12	7.0	15 × 8	Bubble Nebula
12	6946	Cep	ScG	20 34.8	+60 09	9.7	11 × 9	
13	457	Cas	OC	01 19.1	+58 20	6.4	13	Phi Cas Cluster
14	869/884	Per	O double C	02 20.0	+57 08	4.3	30 and 30	Sword Handle
15	6826	Cyg	PN	19 44.8	+50 31	9.8	0.5/2.3	Blinking Nebula
16	7243	Lac	OC	22 15.3	+49 53	6.4	21	
17	147	Cas	dE4G	00 33.2	+48 30	9.3	13 × 8	
18	185	Cas	dE0G	00 39.0	+48 20	9.2	12 × 9	
19	IC 5146	Cyg	BN	21 53.5	+47 16	10.0	12 × 12	Cocoon Neb
20	7000	Cyg	BN	20 58.8	+44 20	6.0	120 × 100	North American Nebula
21	4449	CVn	IG	12 28.2	+44 06	9.4	5 × 3	
22	7662	And	PN	23 25.9	+42 33	9.2	0.3/2.2	
23	891	And	SbG	02 22.6	+42 21	9.9	14 × 2	
24	1275	Per	Seyfert G	03 19.8	+41 31	11.6	2.6 × 1	Per A radio source
25	2419	Lyn	GC	07 38.1	+38 53	10.4	4.1	
26	4244	CVn	SG	12 17.5	+37 49	10.6	16 × 2.5	
27	6888	Cyg	BN	20 12.0	+38 21	7.5	20 × 10	Crescent Nebula
28	752	And	OC	01 57.8	+37 41	5.7	50	
29	5005	CVn	SbG	13 10.9	+37 03	9.8	5.4 × 2	
30	7331	Peg	SbG	22 37.1	+34 25	9.5	11 × 4	
31	IC 405	Aur	BN	05 16.2	+34 16	6.0	30 × 19	Flaming Star Nebula
32	4631	CVn	ScG	12 42.1	+32 32	9.3	15 × 3	
33	6992/5	Cyg	SN	20 56.4	+31 43	–	60 × 8	East Veil Nebula
34	6960	Cyg	SN	20 45.7	+30 43	–	70 × 6	West Veil Nebula
35	4889	Com	E4G	13 00.1	+27 59	11.4	3 × 2	Brightest in cluster
36	4559	Com	ScG	12 36.0	+27 58	9.8	10 × 4	
37	6885	Vul	OC	20 12.0	+26 29	5.7	7	
38	4565	Com	SbG	12 36.3	+25 59	9.6	16 × 3	
39	2392	Gem	PN	07 29.2	+20 55	9.9	0.2/0.7	Eskimo Nebula
40	3626	Leo	SbG	11 20.1	+18 21	10.9	3 × 2	
41	–	Tau	OC	04 27.0	+16 00	1.0	330	Hyades
42	7006	Del	GC	21 01.5	+16 11	10.6	2.8	Very distant globular
43	7814	Peg	SbG	00 03.3	+16 09	10.5	6 × 2	
44	7479	Peg	SBbG	23 04.9	+12 19	11.0	4 × 3	

(continued)

(continued)

C#	NGC/IC	Con.	Type	R. A. h m	Dec. °...′	Mag.	Size ()	Description
45	5248	Boo	ScG	13 37.5	+08 53	10.2	6 × 4	
46	2261	Mon	BN	06 39.2	+08 44	10.0	2 × 1	Hubble's Variable Neb.
47	6934	Del	GC	20 34.2	+07 24	8.9	5.9	
48	2775	Can	SaG	09 10.3	+07 02	10.3	4.5 × 3	
49	2237-9	Mon	BN	06 32.3	+05 03	–	80 × 60	Rosette Nebula
50	2244	Mon	OC	06 32.4	+04 52	4.8	24	
51	IC 1613	Cet	IG	01 04.8	+02 07	9.0	12 × 11	
52	4697	Vir	E4G	12 48.6	−05 48	9.3	6 × 3	
53	3115	Sex	E6G	10 05.2	−07 43	9.1	8 × 3	Spindle Galaxy
54	2506	Mon	OC	08 00.2	−10 47	7.6	7	
55	7009	Aqr	PN	21 04.2	−11 22	8.3	2.5/1	Saturn Nebula
56	246	Cet	PN	00 47.0	−11 53	8.0	3.8	
57	6822	Sgr	IG	19 44.9	−14 48	9.3	10 × 9	Barnard's Galaxy
58	2360	CMa	OC	07 17.8	−15 37	7.2	13	
59	3242	Hya	PN	10 24.8	−18 38	8.6	0.3/21	Ghost of Jupiter
60	4038	Crv	ScG	12 01.9	−18 52	11.3	2.6 × 1.8	The Antennae
61	4039	Crv	ScG	12 01.9	−18 53	13.0	3.2 × 2.2	The Antennae
62	247	Cet	SG	00 47.1	−20 46	8.9	20 × 7	
63	7293	Aqr	PN	22 29.6	−20 48	6.5	13	Helix Nebula
64	2362	CMa	OC	07 18.8	−24 57	4.1	8	Tau CMa Cluster
65	253	Scl	SG	00 47.6	−25 17	7.1	25 × 7	Sculptor Galaxy
66	5694	Hya	GC	14 39.6	−26 32	10.2	3.6	
67	1097	For	SBbG	02 46.3	−30 17	9.2	9 × 6	
68	6729	CrA	BN	19 01.9	−36 57	9.7	1.0	R CrA Nebula
69	6302	Sco	PN	17 13.7	−37 06	12.8	0.8	Bug Nebula
70	300	Scl	SdG	00 54.9	−37 41	8.1	20 × 13	
71	2477	Pup	OC	07 52.3	−38 33	5.8	27	
72	55	Scl	SBG	00 14.9	−39 11	8.2	32 × 6	Brightest in Scl Cluster
73	1851	Col	GC	05 14.1	−40 03	7.3	11	
74	3132	Vel	PN	10 07.7	−40 26	8.2	0.8	
75	6124	Sco	OC	16 25.6	−40 40	5.8	29	
76	6231	Sco	OC	16 54.0	−41 48	2.6	15	
77	5128	Cen	Peculiar Galaxy	13 25.5	−43 01	7.0	18 × 14	Cen A radio source
78	6541	CrA	GC	18 08.0	−43 42	6.6	13	
79	3201	Vel	GC	10 17.6	−46 25	6.7	18	
80	5139	Cen	GC	13 26.8	−47 29	3.6	36	Omega Centauri
81	6352	Ara	GC	17 25.5	−48 25	8.1	7	
82	6193	Ara	OC	16 41.3	−48 46	5.2	15	

(continued)

(continued)

C#	NGC/IC	Con.	Type	R. A. h m	Dec. °…′	Mag.	Size ()	Description
83	4945	Cen	SBcG	13 05.4	−49 28	9.5	20 × 4	
84	5286	Cen	GC	13 46.4	−51 22	7.6	9	
85	IC 2391	Vel	OC	08 40.2	−53 04	2.5	50	o (Omicron) Vel Cluster
86	6397	Ara	GC	17 40.7	−53 40	5.6	26	
87	1261	Hor	GC	03 12.3	−55 13	8.4	7	
88	5823	Cir	OC	15 05.7	−55 36	7.9	10	
89	6087	Nor	OC	16 18.9	−57 54	5.4	12	S Nor Cluster
90	2867	Car	PN	09 21.4	−58 19	9.7	0.2	
91	3532	Car	OC	11 06.4	−58 40	3.0	55	
92	3372	Car	BN	10 43.8	−59 52	6.2	120 × 120	Eta Carinae Nebula
93	6752	Pav	GC	19 10.9	−59 59	5.4	20	
94	4755	Cru	OC	12 53.6	−60 20	4.2	10	Jewel Box Cluster
95	6025	TrA	OC	16 03.7	−60 30	5.1	12	
96	2516	Car	OC	07 58.3	−60 52	3.8	30	
97	3766	Cen	OC	11 36.1	−61 37	5.3	12	
98	4609	Cru	OC	12 42.3	−62 58	6.9	5	
99	–	Cru	DN	12 53.0	−63 00	–	400 × 300	Coal Sack
100	IC 2944	Cen	OC	11 36.6	−63 02	4.5	15	– (Lambda) Cen Cluster
101	6744	Pav	SBbG	19 09.8	−63 51	9.0	16 × 10	
102	IC 2602	Car	OC	10 43.2	−64 24	1.9	50	÷ (Theta) Car Cluster
103	2070	Dor	BN	05 38.7	−69 06	1.0	40 × 25	Tarantula Neb. in LMC
104	362	Tuc	GC	01 03.2	−70 51	6.6	13	
105	4833	Mus	GC	12 59.6	−70 53	7.3	14	
106	104	Tuc	GC	00 24.1	−72 05	4.0	31	47 Tucanae
107	6101	Aps	GC	16 25.8	−72 12	9.3	11	
108	4372	Mus	GC	12 25.8	−72 40	7.8	19	
109	3195	Cha	PN	10 09.5	−80 52	–	0.6	

Key to Object Types: *BN bright nebula, GC globular cluster, OC open cluster, EG elliptical (type) galaxy, DN dark nebula, IG irregular galaxy, PN planetary nebula, SN supernova remnant, SG spiral (type) galaxy*

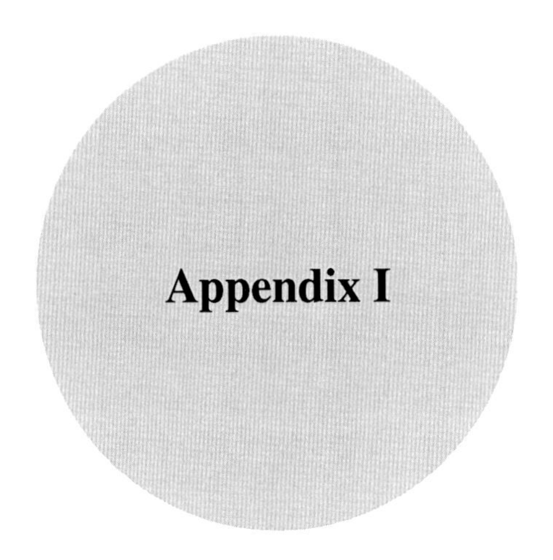

Appendix I

The Herschel 400

Listed here are the Astronomical League's listing of 400 Herschel objects, for which the AL grants the *Herschel Award*. This list was selected and compiled by Brenda F. Guzman (Branchett), Lydel Guzman, Paul Jones, James Morrison, Peggy Taylor, and Sara Saey of the Ancient City Astronomy Club in St. Augustine. It is a subset of William Herschel's work of 2514 deep sky objects. Many of these objects are included in the Messier, Caldwell, and SAA100 catalog.

Many of the objects listed in the Herschel 400 are difficult or not visible visually through the average sized (4" to 8" aperture) size of telescope. It is possible to detect and image the fainter objects through long exposures or multiple exposures and stacking techniques.

J.L. Chen, *Astronomy for Older Eyes*, The Patrick Moore Practical Astronomy Series, DOI 10.1007/978-3-319-52413-9

X	NGC #	H #	Const.	Mag.	Type	Size	Season	RA+Dec	Comment
*	205	18-5	And	8	G	17.4	F	0040.4+4141	M110
	404	224-2	And	10.7	G	2.1 × 2.0	F	0109.4+3543	el circular beautiful brt nucleu
*	752	32-7	And	5.7	OC	50	F	0157.8+3741	70 stars large scattered outward
*	7662	18-4	And	9	PN	32 × 23	F	2325.9+4233	Blue-green oval bright dense
	7686	69-8	And	5.6	OC	7.4	F	2330.2+4908	35 stars loose poor 1 red/orange
*	7009	1-4	Aqr	8.4	PN	44 × 2	F	2104.2−1122	Saturn Nebula
	7606	104-1	Aqr	10.8	G	4.4 × 1.5	F	2319.1−0846	sp elongated elusive large
	7723	110-1	Aqr	11.1	G	2.2 × 1.6	F	2338.9−1258	Bsp faint small elusive
	7727	111-1	Aqr	10.7	G	2.7	F	2339.9−1218	sp elusive slightly elongated
	772	112-1	Ari	10.9	G	5 × 3	F	0159.3+1901	sp elongated elusive
	129	79-8	Cas	6.5	OC	21	F	0029.9+6014	50 stars bright rich 6 mag star
	136	35-6	Cas	11.3	OC	1	F	0031.5+6132	10 stars rich resolvable elongat
	185	707-2	Cas	9.2	G	11.5	F	0039.0+4820	el circular real challenge
	225	78-8	Cas	12	OC	7.5	F	0043.4+6147	21 stars large rich
	278	159-1	Cas	10.9	G	1.3 × 1.3	F	0052.1+4733	el round bright nucleus
	381	64-8	Cas	9.2	OC	6	F	0108.3+6135	24 stars yellow + red-orange star
*	436	45-7	Cas	8.8	OC	4	F	0115.6+5849	25 stars loose poor large brt
*	457	42-1	Cas	6.4	OC	10	F	0119.1+5820	50 stars brt rich scattered bug?
*	559	48-7	Cas	9.5	OC	4	F	0129.5+6318	60 righ bright loosely scattered
*	637	49-7	Cas	8.2	OC	3	F	0142.9+6400	20 stars compact chain of stars
*	654	46-7	Cas	6.5	OC	5	F	0144.1+6153	50 stars loose poor a brt star
*	659	65-8	Cas	7.9	OC	5	F	0144.2+6042	20 stars loose poor near 663
*	663	31-6	Cas	7.1	OC	16	F	0146.0+6115	80 stars rich resolvable nebulous
*	7789	30-6	Cas	6.7	OC	16	F	2357.0+5644	Large rich circular sl. nebulous

(continued)

(continued)

X	NGC #	H #	Const.	Mag.	Type	Size	Season	RA+Dec	Comment
*	7790	56-7	Cas	8.5	OC	17	F	2358.4+6113	25 stars scattered cir nebulous
	40	58-4	Cep	10.2	PN	.6	F	0013.0+7232	Easy round bright greenish
	7142	66-7	Cep	9.3	OC	11	F	2145.9+6548	35 stars rectangular
*	7160	67-8	Cep	6.6	OC	7	F	2153.7+6236	25 bright scattered rich resolved
*	7380	77-8	Cep	7.2	OC	10	F	2247.0+5806	50 stars loose poor
	7510	44-7	Cep	7.9	OC	2	F	2311.5+6034	20 stars loose poor arrowhead
	157	3-2	Cet	10.4	G	2.8 × 2.1	F	0034.8−0824	sp oval 2 stars in it brt nucleus
*	246	25-5	Cet	8.5	PN	3.8	F	0047.0−1153	Large oval difficult 2 field star
	247	20-5	Cet	8.9	G	20	F	0047.1−2046	sp elusive elongated
	584	100-1	Cet	10.8	G	3.8	F	0131.3−0652	el small dim round fair nucleus
	596	4-2	Cet	10.9	G	3.5	F	0132.9−0702	el very small circular
	615	282 2	Cet	11.6	G	4	F	0135.1−0720	sp small elusive
	720	105-1	Cet	10.5	G	4.4	F	0153.0−1344	el small round brt nucleus
	779	101-1	Cet	11.3	G	3.7 × .9	F	0159.7−0558	Bsp edge on extremely elongated
*	7000	37-5	Cyg		DN	120 × 100	F	2058.8+4420	North American Nebula
	7008	192-1	Cyg	13.5	PN	86 × 96	F	2100.6+5433	Seems brighter than 13.5
	7044	24-6	Cyg	11.3	OC	3.5	F	2112.9+4129	40 scattered stars
	7062	51-7	Cyg	8.3	OC	5	F	2123.2+4623	30 stars inter. rich
	7086	32-6	Cyg	8.4	OC	8	F	2130.5+5135	50 stars large scattered
	7128	40-7	Cyg	9.7	OC	2	F	2144.0+5329	20 stars loose poor
	7006	52-1	Del	10.3	GC	1.1	F	2101.5+1611	faint
*	7209	53-7	Lac	6.7	OC	20	F	2205.2+4630	50 stars bright rich resolvable
*	7243	75-8	Lac	6.4	OC	20	F	2215.3+4953	50 stars loose scattered
*	7296	41-7	Lac	9.4	OC	4	F	2228.2+5217	15 stars compact elusive
*	7217	207-2	Peg	10.2	G	2.6 × 2.3	F	2207.9+3122	sp round
	7331	53-1	Peg	9.7	G	10 × 2	F	2237.1+3425	sp faint edge on

(continued)

(continued)

X	NGC #	H #	Const.	Mag.	Type	Size	Season	RA+Dec	Comment
	7448	251-2	Peg	11.2	G	2 × 1	F	2300.1+1559	sp elongated averted vision
	7479	55-1	Peg	11.6	G	3.4 × 2.6	F	2304.9+1219	Bsp slightly elongated elusive
	651	193-1	Per	11	PN	157 × 87	F	0142.3+5135	M76
*	488	252-3	Psc	10.3	G	4.2 × 3.3	F	0121.8+0515	sp circular brt nucleus
*	524	151-1	Psc	10.6	G	3.2	F	0124.8+0932	el small round brt nucleus
*	253	1-5	Scl	7.1	G	25.1	F	0047.6−2517	sp elongated brt center
*	288	20-6	Scl	8.1	GC	13.8	F	0052.8−2635	Large bright nucleus
	613	281-1	Scl	10.2	G	4 × 2	F	0134.3−2925	Bsp slightly elongated
*	598	17-5	Tri	5.7	G	62	F	0133.9+3039	M33
	6755	19-7	Aql	9	OC	10	S	1907.8+0414	35 stars large loose fairly rich
	6756	62-7	Aql	10.7	OC	3	S	1908.7+0441	Very difficult like a globular
	6781	743-3	Aql	11	PN	106	S	1918.4+0633	Circular nebulous no color
*	5466	9-6	Boo	8.5	GC	5	S	1405.5+2832	Large diffuse faint oval elusive
	5557	99-1	Boo	11.6	G	.9 × .8	S	1418.4+3630	el small brt nucleus nebulosity
	5676	189-1	Boo	11.2	G	3 × 1	S	1432.8+4948	sp very faint large oval
	5689	188-1	Boo	11.4	G	2 × .5	S	1435.5+4845	sp faint slight elongated elus
*	6939	42-6	Cep	10	OC	5	S	2031.4+6038	Very rich small appears nebulous
*	6946	76-4	Cep	10.5	G	9 × 7.5	S	2034.8+6009	sp oval large faint wispy disk
*	6826	73-4	Cyg	8.8	PN	22 × 24	S	1944.8+5031	Small fuzzy round some blue
	6834	16-8	Cyg	10.3	OC	4	S	1952.2+2925	15 stars a string scattered star
	6866	59-7	Cyg	9	OC	6	S	2003.7+4400	36 stars rich brt stars large
	6910	56-8	Cyg	7.5	OC	8	S	2023.1+4047	40 stars rich fairly brt large
	6905	16-4	Del	12	PN	44 × 37	S	2022.4+2007	Easy oval silver disk

(continued)

(continued)

X	NGC #	H #	Const.	Mag.	Type	Size	Season	RA+Dec	Comment
*	6934	103-1	Del	10	GC	1.5	S	2034.2+0723	Circular fairly brt nucleus
*	5866	215-1	Dra	10.8	G	2.8 × 1	S	1506.5+5546	el M102
	5907	759-2	Dra	11.3	G	11.1 × .7	S	1515.9+5619	sp needle very long central bulge
	5982	764-2	Dra	10.9	G	1.2 × .8	S	1538.7+5921	el round fuzzy very small
	6543	37-4	Dra	8.8	PN	22	S	1758.6+6638	Oval brt opaque blue-green
*	6207	701-2	Her	11.3	G	2 × 1.1	S	1643.1+3650	sp oval elusive near M13 difficult
	6229	50-4	Her	8.7	GC	1.2	S	1647.0+4732	Faint oval brt central area
	5694	196-2	Hyd	11	GC	2.2	S	1439.6−2632	Small fuzzy smudge brt nucleus
*	5897	19-6	Lib	11	GC	7.3	S	1517.4−2101	Oval partly resolved elusive
*	6171	40-6	Oph	9.2	GC	2.2	S	1632.5−1303	M107
	6235	584-2	Oph	10.4	GC	1.9	S	1653.4−2211	Faint no distinct nucleus diff.
*	6284	11-6	Oph	10.5	GC	1.5	S	1704.5−2446	Easy round fairly bright small
	6287	195-2	Oph	9.9	GC	1.7	S	1705.2−2242	Circular brt nucleus no resolve
*	6293	12-6	Oph	9.5	GC	1.9	S	1710.2−2634	Easy circular small brt nucleus
*	6304	147-1	Oph	9.8	GC	1.6	S	1745.5−2928	Circular brt nucleus no resolve
	6316	45-1	Oph	10.0	GC	1.1	S	1716.6−2808	Oval small bright
*	6342	149-1	Oph	10.0	GC	.5	S	1721.2−1935	Small faint central area brt
	6355	46-1	Oph	10.5	GC	1	S	1724.0−2621	Circular elus extremely faint
*	6356	48-1	Oph	9.5	GC	1.7	S	1723.6−1749	Fairly large brt nucleus no resolved
*	6369	11-4	Oph	9.9	PN	28	S	1729.3−2346	Round blue-green Pipe Nebula
	6401	44-1	Oph	11	GC	1	S	1738.6−2355	brt circular no resolution
	6426	587-2	Oph	11.5	GC	1.3	S	1744.9+0300	Faint elus resembles a nebula
	6517	199-2	Oph	10.5	GC	.4	S	1801.8−0858	Faint elus brt central area

(continued)

(continued)

X	NGC #	H #	Const.	Mag.	Type	Size	Season	RA+Dec	Comment
*	6633	72-8	Oph	5.5	OC	20	S	1827.7+0634	44 stars rich resolved. A red star
	6144	10-6	Sco	10.5	GC	3.3	S	1627.3−2602	Faint oval fairly large elusive
*	6664	12-8	Sct	8.9	OC	18	S	1836.7−0813	25 stars poor loose scattered
*	6712	47-1	Sct	10	GC	2.1	S	1853.1−0842	Round opaque touches PN IC1298
	6118	402-2	Ser	11.5	G	4.3 × 1.3	S	1621.8−0217	sp Blinking Galaxy elus near mag 6
	6440	150-1	Sgr	10.4	GC	.7	S	1748.9−2022	brt nucleus no resolution
	6445	586-2	Sgr	11	PN	38 × 29	S	1749.2−2001	Circular brt blue-green
	6451	13-6	Sgr	8.5	OC	6	S	1750.7−3013	Tightly grouped large faint in
*	6514	41-1	Sgr	6.9	OC	29 × 27	S	1802.3−2302	M20 Trifid Nebula star cluster
	6520	7-7	Sgr	8.1	OC	5	S	1803.4−2754	25 stars easily resolvable
	6522	49-1	Sgr	9.5	GC	.7	S	1803.6−3002	Faint small some resolution
	6528	200-2	Sgr	10.5	GC	.5	S	1804.8−3003	Near 6522 some resolution
	6540	198-2	Sgr	11	OC	.5	S	1806.3−2749	20 stars rich small diffuse
	6544	197-2	Sgr	10	GC	1	S	1807.3−2500	brt oval resolution good edges
	6553	12-4	Sgr	10	GC	1.7	S	1809.3−2554	Righ tight fuzzy resolution
	6568	30-7	Sgr	8.5	OC		S	1812.8−2136	33 stars loose poor wide
	6569	201-2	Sgr	10.4	GC	1.4	S	1813.6−3150	Easy faint brt central area
	6583	31-7	Sgr	11.5	OC	1.5	S	1815.8−2208	Elus faint rich like a globular
	6624	50-1	Sgr	9.5	GC	2	S	1823.7−3022	Fairly large brt nucleus
	6629	204-2	Sgr	10.6	PN	16 × 14	S	1825.7−2312	Circular no color brt nucleus
*	6638	51-1	Sgr	10.2	GC	1.4	S	1830.9−2530	Small faint elusive
	6642	205-2	Sgr	10.5	GC	1	S	1831.9−2329	Very faint small elusive
	6645	23-6	Sgr	8.5	OC	10	S	1832.6−1645	50 stars rich compressed brt

(continued)

(continued)

X	NGC #	H #	Const.	Mag.	Type	Size	Season	RA+Dec	Comment
*	6818	51-4	Sgr	10	PN	22 × 15	S	1944.0−1409	Diffuse fuzzy rount some blue
	5473	231-1	UMa	11.4	G	.9 × .7	S	1404.7+5454	el companion to M101 round elus.
*	5474	214-1	UMa	11.4	G	4 × 2.9	S	1405.0+5340	sp near M101 dim diffuse elusive
	5631	236-1	UMa	11.4	G	.7 × .7	S	1426.6+5635	sp small round faint difficult
	6217	280-1	UMn	11.5	G	1.8 × 1.2	S	1632.6+7812	sp faint sl oval elusive arms
	5566	144-1	Vir	10.4	G	5.6 × 1.1	S	1420.3+0356	sp edgeon faint lg brt nuc elus
	5576	146-1	Vir	11.7	G	1 × .8	S	1421.1+0316	el small faint brt nuc nebulosity
*	5634	70-1	Vir	10.4	GC	1.3	S	1429.6−0559	Small fuzzy patch opaque oval
	5746	126-1	Vir	10.1	G	6.2 × .8	S	1444.9+0157	sp edge-on needle faint large elus
	5846	128-1	Vir	10.5	G	.9 × .9	S	1506.4+0136	el small circular fairly bright
	6802	14-6	Vul	11	OC	3.5	S	1930.6+2016	Faint small opaque disk of stars
*	6823	18-7	Vul	9.8	OC	5	S	1943.1+2318	30 stars some nebulosity
*	6830	9-7	Vul	9	OC	8	S	1951.0−2304	20 stars brt large easy
	6882	22-8	Vul	5.5	OC		S	2011.7+2633	20 stars small in a rich field
	6885	20-8	Vul	9.1	OC	20	S	2012.0+2629	42 stars large a brt star rich
*	6940	8-7	Vul	6.5	OC	10	S	2034.6+2818	100 stars very brt rich large
	5248	34-1	Boo	11.3	G	6.1 × 4.4	Sp	1337.5+0853	sp lg oval brt nucleus av
	2655	288-1	Cam	10.7	G	5 × 3.4	Sp	0855.6+7813	sp faint face-on low surf brt
	2775	2-1	Cnc	10.7	G	2.3 × 1.9	Sp	0910.3+0702	sp circular brt nucleus
	4147	19-1	Com	9.4	GC	1.7	Sp	1210.1+1833	Small starlike outer nebulosity
	4150	73-1	Com	11.6	G	1.2 × .9	Sp	1210.6+3024	el round starlike nucl a brt sta
	4203	175-1	Com	11	G	1.8 × 1.5	Sp	1215.1+3312	el sl oval faint brt nucleus av
	4245	74-1	Com	11.1	G	1.5 × 1	Sp	1217.6+2936	sp elus starlike hint elong av

(continued)

(continued)

X	NGC #	H #	Const.	Mag.	Type	Size	Season	RA+Dec	Comment
	4251	89-1	Com	10.2	G	2.3 × .5	Sp	1218.1+2810	sp round starlike nebulosity
	4274	75-1	Com	10.8	G	6.7 × 1.3	Sp	1219.8+2937	sp lg fuzzy brt nucl elongated
	4278	90-1	Com	10.3	G	1.4 × 1.3	Sp	1220.1+2917	el starlike brt nucl defined
	4293	5-5	Com	11.5	G	4.6 × 1.6	Sp	1221.2+1823	pec edge-on granulated elus elon
	4314	76-1	Com	10.8	G	3.1 × 2.9	Sp	1222.6+2953	Bsp oval diffuse brter nucleus
	4350	86-2	Com	11	G	1.8 × .5	Sp	1224.0+1642	el round starlike nebulosity
	4394	55-2	Com	11.2	G	2.3 × 2.3	Sp	1225.9+1813	Bsp round brt nuc av near M85
	4414	77-1	Com	9.7	G	3.2 × 1.5	Sp	1226.4+3113	sp starlike nuc extended arms
	4419	113-1	Com	11.4	G	2.2 × .6	Sp	1226.9+1503	el elong brt nucl fainter arm av
	4448	91-1	Com	11.4	G	2.8 × 1	Sp	1228.2+2837	sp sl elong uniform av elus
*	4450	56-2	Com	10	G	3 × 2.5	Sp	1228.5+1705	sp round brt distinct nucleus
	4459	161-1	Com	10.9	G	1.2 × 1	Sp	1229.0+1359	el brt sl elong
*	4473	114-2	Com	10.1	G	1.6 × .9	Sp	1229.8+1326	el round starlike nucl brt
*	4477	115-2	Com	10.7	G	2.4 × 2.2	Sp	1230.0+1438	sp round starlike nucl fair brt
*	4494	83-1	Com	9.6	G	1.3 × 1.4	Sp	1231.4+2547	el brt starlike neb round small
*	4548	120-2	Com	10.8	G	3.7 × 3.2	Sp	1235.4+1430	Bsp M91
*	4559	92-1	Com	10.6	G	11 × 4.5	Sp	1236.0+2758	sp elong uniform fairly brt
*	4565	24-5	Com	10.2	G	14.4 × 1.2	Sp	1236.3+2559	sp uniform needle beautiful
	4689	128-2	Com	11.5	G		Sp	1247.8+1346	sp round uniform av
*	4725	84-1	Com	8.9	G	10 × 5.5	Sp	1250.4+2530	sp oval starlike nucl easy
	4027	296-2	Cor	11.5	G	2.4 × 2	Sp	1159.5−1916	sp faint elus hint elong av
	4038	28.1-4	Cor	11.5	G	2.5 × 2.5	Sp	1201.9−1852	sp round brt nucleus
*	4361	65-1	Cor	10.8	PN	81	Sp	1224.5−1848	lg round fuzzy no color noted

(continued)

(continued)

X	NGC #	H #	Const.	Mag.	Type	Size	Season	RA+Dec	Comment
	3962	67-1	Cra	11.3	G	1.1 × .9	Sp	1154.7−1358	el round very faint brt nucleus
	4111	195-1	CVn	9.7	G	3.3 × .6	Sp	1207.1+4304	el elong starlike outer arms av
	4143	54-4	CVn	11	G	1.4 × .9	Sp	1209.6+4232	el round starlike nucleus small
	4151	165-1	CVn	11.6	G	2.5 × 1.6	Sp	1210.5+3924	pec uniform round ovalish elus
	4214	95-1	CVn	10.3	G	6.6 × 5.8	Sp	1215.6+3620	irr very brt large diffuse oval
*	4258	43-55	CVn	8.6	G	19.5 × 7	Sp	1219.0+4718	sp M106
	4346	210-1	CVn	11.6	G	1.9 × .7	Sp	1223.5+4700	el round almost starlike neb
*	4449	213-1	CVn	9.2	G	4.1 × 3.4	Sp	1228.2+4406	irr sl elong uniform pretty
	4485	197-1	CVn	11.6	G	1.5 × .8	Sp	1230.5+4142	irr uniform brt elusive small
*	4490	198-1	CVn	9.7	G	5.6 × 2.1	Sp	1230.6+4138	sp lg elong uniform brt easy
*	4618	178-1	CVn	11.7	G	.5 × 3	Sp	1241.5+4109	sp lg sl elong uniform brt
	4631	42-4	CVn	9.3	G	12.6 × 1.4	Sp	1242.1+3232	sp elong uniform brt easy
	4656	176-1	CVn	11.2	G	18 × 2	Sp	1244.0+3210	pec lg elus needle av
	4800	211-1	CVn	11.1	G	1.2 × 1	Sp	1254.6+4642	sp round brt stands out well
*	5005	96-1	CVn	9.8	G	4.4 × 1.7	Sp	1310.9+3703	sp elong lg brt nucleus
*	5033	97-1	CVn	10.3	G	9.9 × 4.8	Sp	1313.4+3636	sp elong brt needle lg
*	5195	186-1	CVn	8.4	G	2 × 1.5	Sp	1330.0+4716	pec companion to M51 round
	5273	98-1	CVn	11.5	G	.9 × .8	Sp	1342.1+3539	el brt nucleus av stands out wel
	3147	79-1	Dra	10.9	G	3 × 2.3	Sp	1016.9+7339	sp lg circular near 6 mag star
*	2548	22-6	Hyd	5.3	OC	30	Sp	0813.8−0548	M48
	2811	505-2	Hyd	11.7	G	1.6 × .5	Sp	0916.2−1606	sp uniform brightness
*	3242	27-4	Hyd	9	PN	40 × 34	Sp	1024.8+1838	nice disk blue-green fairlylarge
	3621	241-1	Hyd	10.5	G	5 × 2	Sp	1118.3+3249	sp lg round near some field star
	3686	160-2	Leo	11.4	G	2.4 × 1.8	Sp	1127.7+1713	sp strlike brt nucleus nebulosity

(continued)

(continued)

X	NGC #	H #	Const.	Mag.	Type	Size	Season	RA+Dec	Comment
*	2903	56-1	Leo	9.1	G	11 × 4.6	Sp	0932.2−2130	sp lg elong needle easy
	2964	114-1	Leo	11	G	2.2 × 1.1	Sp	0942.9+3151	sp round brt nucleus faint rim
	3190	44-2	Leo	11.3	G	3 × 1	Sp	1018.1+2150	sp fuzzy round same field 3193
	3193	45-2	Leo	11.5	G	.9 × .0	Sp	1018.4+2154	el fuzzy roundish uniform c 3190
	3226	28-22	Leo	11.5	G	1 × .8	Sp	1023.4+1954	el round nebulous uniform
	3227	29-2	Leo	11.4	G	3 × 1.2	Sp	1023.5+1952	sp round nebulous near NGC 3226
	3377	99-2	Leo	10.5	G	1.9 × 1	Sp	1047.7+1359	el small brt nucleus
*	3379	17-1	Leo	9.5	G	2.2 × 2	Sp	1047.8+1235	el M105
	3384	18-1	Leo	10.2	G	4.4 × 1.4	Sp	1048.3+1238	el round brt nucleus >NGC 3379
	3395	116-1	Leo	12	G	1.5 × .9	Sp	1049.8+3259	sp elus sl elong faint nucleus
	3412	27-1	Leo	10.4	G	2.4 × 1.1	Sp	1050.9+1325	el circular very difficult
	3489	101-2	Leo	11.5	G	2 × .9	Sp	1100.3+1354	el round impressive brt nucleus
*	3521	13-1	Leo	10.5	G	7 × 4	Sp	1105.8−0002	sp elongated very impressive
	3593	29-1	Leo	11.3	G	2.5 × .9	Sp	1114.6+1249	sp round faint inconspicuous
*	3607	50-2	Leo	9.6	G	1.7 × 1.5	Sp	1116.9+1803	el like a hairy star round brt
	3608	51-2	Leo	11.1	G	1.4 × 1	Sp	1117.0+1809	el round small same field 3607
*	3626	52-2	Leo	10.5	G	1.6 × 1.1	Sp	1120.1+1821	sp starlike brt nucleus v small
*	3628	8-5	Leo	10.9	G	12 × 1.5	Sp	1120.3+1336	sp edge-on large beautiful
	3640	33-2	Leo	10.7	G	1.1 × 1	Sp	1121.1+0314	el round small faint near a star
	3810	21-1	Leo	10.8	G	3.6 × 2.5	Sp	1141.0+1128	sp sl lg brt nucleus
	3900	82-1	Leo	11.5	G	1.7 × .8	Sp	1149.2+2701	sp starlike sl neb near 3912
	3912	342-2	Leo	11.5	G	.9 × .5	Sp	1150.0+2629	sp small round uniform near 3900
	2859	137-1	LMn	10.7	G	4.4 × 3.5	Sp	0924.3+3431	Bsp brt nucleus fainter arm
	3245	86-1	LMn	11.2	G	1.8 × .9	Sp	1027.3+2830	el round fuzzy uniform

(continued)

(continued)

X	NGC #	H #	Const.	Mag.	Type	Size	Season	RA+Dec	Comment
	3277	359-2	LMn	12	G	1.1 × .9	Sp	1032.9+2831	sp round brt nucleus
	3294	164-1	LMn	11.4	G	2.6 × 1.2	Sp	1036.3+3720	sp very difficult uniform brt
	3344	81-1	LMn	11	G	7.6 × 6.2	Sp	1043.5+2455	sp round faint near mag 9 star
	3414	362-2	LMn	11	G	1.4 × 1	Sp	1051.3+2759	Bsp starlike fuzzy nebulosity
	3432	172-1	LMn	11.4	G	5.8 × .8	Sp	1052.5+3637	sp elongated uniform brt 2 stars
	3486	87-1	LMn	11	G	6.8 × 4.5	Sp	1000.4+2858	sp round faint fuzzy uniform
	3504	88-1	LMn	10.9	G	2.2 × 2.2	Sp	1100.5+2758	sp round very impressive
*	2683	200-1	Lyx	9.6	G	8 × 13	Sp	0852.7+3325	sp edge-on pretty needle
	2782	167-1	Lyx	11.7	G	1.8 × 1.6	Sp	0914.1+4007	sp starlike some nebulosity
	2527	30-8	Pup	8	OC	22	Sp	0805.3−2810	50 stars loose scattered
*	2539	11-7	Pup	8.2	OC	21	Sp	0810.7−1250	a brt center blue one 19 Pup
	2567	64-7	Pup	8.3	OC	10	Sp	0818.6−3038	50 stars brt + faint ones
	2571	39-6	Pup	7.5	OC	8	Sp	0818.9−2944	25 stars loose irregular faint
	2613	266-2	Pyx	11	G	6.6 × 1.3	Sp	0833.4−2258	sp edge-on elus difficult
	2627	63-7	Pyx	8.3	OC	8	Sp	0837.3−2957	40 stars small hard distinguish
	2974	61-1	Sex	11	G	1.5 × .9	Sp	0942.6−0342	sp elong fuzzy low surf brt
*	3115	163-1	Sex	9.3	G	4 × 1.2	Sp	1005.2+0743	el small circular
	3166	3-1	Sex	11.4	G	4.4 × 1.7	Sp	1013.8+0326	sp brt nucleus round field 3169
	3169	4-1	Sex	11.7	G	4 × 1.7	Sp	1014.2+0328	sp brt nucleus sl round
*	2681	242-1	UMa	10.4	G	2.8 × 2.5	Sp	0853.5+5119	sp circular faint fuzzy
	2742	249-1	UMa	11.2	G	2.5 × 1	Sp	0907.6+6029	sp elongated elus
	2768	250-1	UMa	10.5	G	2 × 1	Sp	0911.6+6002	el sl elong same field NGC 2742
	2787	216-1	UMa	10.9	G	2 × 1.3	Sp	0919.3+6912	sp elong near 2 brt stars

(continued)

(continued)

X	NGC #	H #	Const.	Mag.	Type	Size	Season	RA+Dec	Comment
*	2841	205-1	UMa	9.3	G	6.4 × 2.4	Sp	0922.0+5058	sp brt elong easy
	2950	68-4	UMa	10.9	G	1.3 × .9	Sp	0942.6+5851	sp starlike fuzzy nebula on edges
	2976	285-1	UMa	8.5	OC	10	Sp	0947.3+6755	50 stars rich concentrate nebulosity
	2985	78-1	UMa	10.6	G	5.5 × 5	Sp	0950.4+7217	sp brt nucleus easy
*	3034	79-4	UMa	8.8	G	9 × 4	Sp	0955.8+6941	pec M82
*	3077	286-1	UMa	10.9	G	2.3 × 1.9	Sp	1003.3+6844	el faint low surf brt comp. M81
	3079	47-5	UMa	11.2	G	8 × 1	Sp	1002.0+5541	sp edge-on faint difficult
*	3184	168-1	UMa	9.6	G	5.6 × 5.6	Sp	1018.3+4125	sp low surf brt difficult round
	3198	199-1	UMa	11	G	9 × 3.2	Sp	1019.9+4533	sp elong with 3 brt stars
*	3310	60-4	UMa	10.1	G	4 × 3	Sp	1035.7+5330	irr round brt nucleus
*	3556	46-5	UMa	11	G	7.7 × 1.3	Sp	1111.5+55.4	sp M108 edge-on
	3610	270-1	UMa	11.2	G	1.3 × 1	Sp	1118.4+5847	el starlike fuzzy nebulous av
	3613	271-1	UMa	11.2	G	1.6 × .8	Sp	1118.6+5800	el starlike like a hairy star
	3619	244-1	UMa	11.7	G	1 × 1	Sp	1119.4+5646	sp starlike same field as 3613
	3631	226-1	UMa	11.2	G	4.3 × 3.2	Sp	1121.0+5310	sp face-on faint uniform fuzzy
	3655	5-1	UMa	11.3	G	1.2 × .9	Sp	1122.9+1635	sp small fuzzy round faint
	3665	219-1	UMa	11.4	G	1.6 × 1.2	Sp	1124.7+3846	el round brt nucleus small
*	3675	194-1	UMa	11.5	G	4 × 1.7	Sp	1126.1+4335	sp faint fuzzy elongation
*	3726	730-2	UMa	10.8	G	5.7 × 3.4	Sp	1133.3+4702	sp round uniform largish
	3729	222-1	UMa	11.7	G	1.8 × 1.3	Sp	1133.8+5308	pec starlike nebulosity
	3813	94-1	UMa	11.7	G	1.7 × .8	Sp	1141.3+3633	sp oval well defined 1 side brt
	3877	201-1	UMa	10.9	G	4.4 × .8	Sp	1146.1+4730	sp elongated uniform brightness
*	3893	738-2	UMa	11.3	G	3.7 × 1.9	Sp	1148.6+4843	sp sl elong brt nucleus av
	3898	228-1	UMa	11.5	G	2.6 × 1	Sp	1149.2+5606	sp round brt nucleus

(continued)

(continued)

X	NGC #	H #	Const.	Mag.	Type	Size	Season	RA+Dec	Comment
	3938	203-1	UMa	11.5	G	4.5 × 3.8	Sp	1152.8+4407	sp lg elong uniform
	3941	173-1	UMa	9.8	G	1.8 × 1.2	Sp	1152.9+3659	sp round nucleus nebulosity
	3945	251-1	UMa	10.8	G	5.2 × 2.2	Sp	1153.2+6041	Bsp diffuse strlike lg elongated
*	3949	202-1	UMa	11	G	2.3 × 1.1	Sp	1153.7+4752	sp oval diffuse uniform brtness
*	3953	45-5	UMa	10.7	G	5.6 × 2.3	Sp	1153.8+5220	sp round brt nucleus av
	3982	62-4	UMa	11.3	G	1.7 × 1.3	Sp	1156.5+5508	sp round elus av brt nucleus
*	3992	61-4	UMa	10.8	G	6.2 × 3.5	Sp	1157.6+5323	sp M109
	3998	229-1	UMa	11.3	G	3.7 × 1.9	Sp	1157.9+5527	el brt nucelus elong elus av
*	4026	223-1	UMa	10.7	G	3.6 × .7	Sp	1156.9+5058	el round starlike nebulosity
	4036	253-1	UMa	10.7	G	2.4 × .9	Sp	1201.4+6153	el oval brt nucleus small
	4041	252-1	UMa	11	G	2.4 × 1.8	Sp	1202.2+6208	sp round brt nucleus av
	4051	56-4	UMa	11	G	4.5 × 3.6	Sp	1203.2+4432	sp lg elus nebulosity av
	4085	224-1	UMa	11.8	G	2.2 × .5	Sp	1205.4+5012	sp brt starlike nucleus oval
	4088	206-1	UMa	10.9	G	4.5 × 1.4	Sp	1205.6+5033	sp elong brt elusive large
	4102	225-1	UMa	11.8	G	2.2 × 1	Sp	1206.4+5243	sp hint of elong starlike brt nu
	5322	256-1	UMa	10	G	1 × 4 × 1.4	Sp	1349.3+6012	el round brt nucleus
	4030	121-1	Vir	11	G	3.1 × 2.2	Sp	1200.4−0106	sp round brt pretty 2 field star
	4179	9-1	Vir	11.6	G	2.7 × .6	Sp	1212.9+0118	el edge-on well distinct arms
*	4216	35-1	Vir	10.4	G	7.4 × .9	Sp	1215.9+1309	sp edge-on needle easy brt nucleus
*	4261	139-2	Vir	10.3	G	.9 × .7	Sp	1219.4+0549	el round uniform fuzzy elus
	4273	569-2	Vir	11.6	G	1.5 × 1	Sp	1219.9+0521	sp starlike near group stars
	4281	573-2	Vir	11.3	G	1.1 × .6	Sp	1220.4+0523	el elus starlike hint neb av
*	4303	139-1	Vir	10.1	G	5.6 × 5.3	Sp	1219.4+0428	sp M61
	4365	30-1	Vir	11.1	G	1.3 × 1	Sp	1224.5+0719	el round faint nucleus brter

(continued)

(continued)

X	NGC #	H #	Const.	Mag.	Type	Size	Season	RA+Dec	Comment
	4371	22-1	Vir	11.6	G	2.2 × 1.2	Sp	1224.9+1142	Bsp brt lg oval opaque brt nucleus
	4429	65-2	Vir	11.2	G	3.3 × 1	Sp	1227.4+1107	sp oval nucleus dominated
*	4435	28.1-1	Vir	10.3	G	1.3 × .8	Sp	1227.7+1305	el elus uniform near 4438 av
*	4438	28.2-1	Vir	10.8	G	8.9 × 3	Sp	1227.8+1301	sp elong uniform lg av
*	4442	156-2	Vir	10.8	G	1.8 × .9	Sp	1228.1+0948	el ovalish starlike neb small
	4478	124-2	Vir	10.9	G	.8 × .7	Sp	1230.3+1220	el round starlike elus neb small
	4526	31-1	Vir	10.9	G	3.3 × 1	Sp	1234.0+0742	el oval faint neb near 2 7mag st
	4527	37-2	Vir	11.5	G	5.3 × 1	Sp	1234.1+0239	sp elong uniform brightness
*	4535	500-2	Vir	11	G	6 × 4	Sp	1234.3+0812	sp elus elong brt nucleus
	4536	2-5	Vir	10.9	G	6.9 × 2.6	Sp	1234.5+0211	sp elong only can see part of it
	4546	160-1	Vir	10	G	1.8 × .8	Sp	1235.5−0348	el round nebulosity uniform
	4550	36-1	Vir	11.7	G	1.4 × .4	Sp	1235.5+1213	el round brt nucleus elusive
	4570	32-1	Vir	10.9	G	2.4 × .5	Sp	1236.9+0715	el fuzzy round
*	4594	43-1	Vir	8.7	G	6 × 2.5	Sp	1240.0−1137	sp M104 The Sombrero Hat
	4596	24-1	Vir	11.4	G	1.8 × 2.2	Sp	1239.9+1011	Bsp round starlike nucleus
*	4636	38-2	Vir	10.4	G	1.4 × 1.3	Sp	1242.8+0241	el round brt nucleus
*	4643	10-1	Vir	10.6	G	1.65 × .9	Sp	1243.3+0159	Bsp round starlike brt nucleus
	4654	126-2	Vir	11	G	4.2 × 2.2	Sp	1244.0+1308	sp lg elong stands out well
	4660	71-2	Vir	10.9	G	1.5 × .8	Sp	1244.5+1111	el round starlike nebulosity
*	4665	142-1	Vir	11.1	G	3.1 × 2.1	Sp	1245.1+0303	Bsp almost stellar small
	4666	15-1	Vir	11.4	G	3.8 × .8	Sp	1245.1−0028	sp lg brt elong stands out well
*	4697	39-1	Vir	10.5	G	2.2 × 1.4	Sp	1248.6−0548	el brt nucleus lg uniform
	4698	8-1	Vir	11.3	G	3 × 1.1	Sp	1248.4+0829	sp sl elong near 2 field stars

(continued)

(continued)

X	NGC #	H #	Const.	Mag.	Type	Size	Season	RA+Dec	Comment
*	4699	129-1	Vir	9.3	G	3 × 2	Sp	1249.0−0840	sp lg easy brt nucleus
	4753	16-1	Vir	10.8	G	3.3 × 1.1	Sp	1252.4−0112	sp brt elong uniform brt
*	4754	25-1	Vir	10.5	G	1 × 1.2	Sp	1252.3+1119	el round brt nucl neb uniform
	4762	75-2	Vir	11	G	3.7 × .4	Sp	1252.9+1114	sp elong uniform lg 3 field star
*	4781	134-1	Vir	11.2	G	2.3 × 1.1	Sp	1254.4−1032	sp uniform elong elus
	4845	536-2	Vir	11.5	G	4.2 × .7	Sp	1258.0+0135	sp elong uniform brt easy
	4856	68-1	Vir	11.5	G	2 × .7	Sp	1259.3−1502	eloval brt nucl easy to see
	4866	162-1	Vir	11.4	G	6.8 × .8	Sp	1259.5+1410	sp elongated needle impress easy
	4900	143-1	Vir	11.3	G	1.7 × 1.5	Sp	1300.6+0230	sp starlike center oval small
	4958	130-1	Vir	10.9	G	1.7 × .7	Sp	1305.8−0801	el round starlike uniform brt
	4995	42-1	Vir	11.2	G	2 × 1.1	Sp	1309.7−0750	sp sl elong elus uniform av
	5054	513-2	Vir	11.5	G	3.8 × 2.2	Sp	1317.0−1638	sp elongated cluster uniform av
	5363	6-1	Vir	10.7	G	1 × 1.4	Sp	1356.1+0529	wl round brt nucl stands out wel
	5364	534-2	Vir	11	G	6.2 × 3	Sp	1356.2+0501	sp lg elongated faint av
*	891	19-5	And	11.5	G	11.8 × 1.1		W0222.6+4221	sp extremely elongated edge-on
	1664	59-8	Aur	7.5	OC	15	W	0451.1+4342	25 stars faint rich elongated
	1857	33-7	Aur	8.5	OC	9	W	0520.2+3921	30 stars no shape poor loose
	1907	39-7	Aur	9.9	OC	5	W	0528.0+3519	Tight some nebulosity richish
	1931	261-1	Aur	9.5	DN	3 × 3	W	0531.4+3519	Starlike with nebulosity around
	2126	68-8	Aur	10	OC	6.5	W	0603.0+4954	28 stars faint poor blue white
	2281	71-8	Aur	6.9	OC	17	W	0649.3+4104	30 stars bright lg easy
*	1501	53-4	Cam	13.3, 9	PN		W	0407.0+6055	Round distict blue-green
*	1502	47-7	Cam	5.3	OC	7	W	0407.7+6220	15 stars compact cross-bow doubl

(continued)

(continued)

X	NGC #	H #	Const.	Mag.	Type	Size	Season	RA+Dec	Comment
	1961	747-3	Cam	11.7	G	3.7 × 1.6	W	0542.1+6923	sp faint elus fuzzy sl elongated
	2403	44-5	Cam	8.9	G	16.8 × 10	W	0736.9+6536	sp lg sl elon face on faint star
	1027	66-8	Cas	7.5	OC	7	W	0242.7+6133	Large loose near 7 mag star
	908	153-1	Cet	11	G	5 × 2.3	W	0223.1−2114	sp elongated evenly brt large
	936	23-4	Cet	10.7	G	3.3 × 2.5	W	0227.6−0109	sp round starlike nucleus
	1022	102-1	Cet	11.2	G	1.8 × 1.4	W	0238.5−0640	Bsp small almost stellar faint
	1052	63-1	Cet	11.2	G	1.3 × 1	W	0241.1−0815	el stellar small hint nebulosity
*	1055	1-1	Cet	11.5	G	6.7 × 1.5	W	0241.8+0026	sp edge on elus same field M77
	2204	13-7	CMj	9.1	OC	13	W	0615.7−1839	scattered a yellow-orange star
	2354	16-7	CMj	9	OC	25	W	0714.3−2544	lg dim scattered round
*	2360	12-7	CMj	9.4	OC	12	W	0717.8−1537	pretty brt no shape near mag 6
	2362	17-7	CMj	10.5	OC	6	W	0718.8−2457	40 stars lg brt in center temple
	1084	64-1	Eri	11	G	2.1 × 1.1	W	0243.0−0735	Easy round brt nucleus
*	1407	107-1	Eri	10.6	G	1.1 × 1.1	W	0340.2−1835	el round starlike brt nucleus
*	1535	26-4	Eri	9.3	PN	20 × 17	W	0414.2−1244	Small round starlike blue tint
*	2129	26-8	Gem	7.2	OC	5	W	0601.0+2318	13 stars broken chainlike
	2158	17-6	Gem	11	OC	4	W	0607.5+2406	brt lg same field as M35
*	2266	21-6	Gem	9.8	OC	5	W	0643.2+2658	30 stars 1 8mag triangular shape
	2304	2-6	Gem	10.1	OC	5.5	W	0655.0+1801	Small loose irregular
*	2355	6-6	Gem	9.5	OC	9	W	0716.9+1347	70 stars tight faint near mag 6
	2371	316-2	Gem	11	PN	54 × 35	W	0725.6+2929	Elongated off-white part of 2372
	2372	317-2	Gem	9.5	PN	47 × 43	W	0725.6+2930	Elongated off-white part ot 2371

(continued)

(continued)

X	NGC #	H #	Const.	Mag.	Type	Size	Season	RA+Dec	Comment
*	2392	45-4	Gem	9.35	PN	47 × 43	W	0729.2+2055	Small round bluish opaque
*	2395	11-8	Gem	9.4	OC	12	W	0727.1+1335	Faint elus rich scattered
*	2420	1-6	Gem	10.2	OC	7	W	0738.5+2134	20 stars some nebulosity
*	1964	21-4	Lep	11.6	G	5.4 × 1.1	W	0533.4−2157	sp sl elongation nucleus visible
	2419	218-1	Lyx	11.5	GC	1.7	W	0738.1+3853	Small dim stands out well
	2215	20-7	Mon	8.6	OC	8	W	0621.0−0717	10 stars round faint a brt star
	2232	25-8	Mon	4	OC	20	W	0626.6−0445	Naked eye brt stars one blue-white
*	2244	2-7	Mon	6.2	OC	24	W	0632.4+0452	16 stars within Rosette Nebula
	2251	3-8	Mon	8.5	OC	10	W	0634.7+0822	30 stars lg flat in appearance
*	2264	27-5,5-8	Mon	4.7	OC D	30	W	0641.1+0953	20 stars loose poor DN nebulosity
	2286	31-8	Mon	8	OC	15	W	0647.6−0310	50 stars loose scattered
	2301	27-6	Mon	5.8	OC	15	W	0651.8+0028	40 stars large pretty easy
	2311	60-8	Mon	9.6	OC	7	W	0657.8−0435	25 stars brt rich easy resolvable
*	2335	32-8	Mon	9.1	Oc	12	W	0706.6−1005	35 stars triangular tight group
	2343	33-8	Mon	8	OC	7	W	0708.3−1039	15 stars brt rich tight cute
*	2353	34-8	Mon	5.3	OC	20	W	0714.6−1018	25 stars loose 1 bl/wh easy
	2506	37-6	Mon	8.5	OC	10	W	0800.2−1047	50 stars rich nebulousity easy
	1788	32-5	Ori	11	DN	8 × 5	W	0506.9−0321	Round non uniform near 8 magstar
	1980	31-5	Ori		DN	14 × 14	W	0535.4−0554	Surround iota Orionis nebulosity
	1999	33-4	Ori	10	DN	16 × 12	W	0536.5−0642	brt nucleus hazy lg difficult
	2022	34-4	Ori	11.5	PN	28 × 27	W	0542.1+0905	Fuzzy oval small slight blue-gr
	2024	28-5	Ori	10.7	DN	30 × 30	W	0541.9−0151	Irregular patchy brt very large
*	2169	24-8	Ori	6.4	OC	5	W	0608.4+1357	15 stars brt tightly grouped

(continued)

X	NGC #	H #	Const.	Mag.	Type	Size	Season	RA+Dec	Comment
	2185	20-4	Ori	11	DN	2 × 2	W	0611.1−0613	brt fuzzy with NGC 2183 + 2184
	2186	25-7	Ori	9.5	OC	5	W	0612.2+0527	30 stars loose poor some brt
*	2194	5-6	Ori	9.2	OC	8	W	0613.8+1248	rich slightly circular resolve
*	869	33-6	Per	4.4	OC	36	W	0219.0+5709	Double Cluster resolvable
*	884	34-6	Per	4.7	OC	36	W	0222.4+5707	Double Cluster resolvable
*	1023	156-1	Per	10.5	G	4 × 1.2	W	0240.3+3904	el elonated brt starlike nucl
*	1245	25-6	Per	6.9	OC	10	W	0314.7+4715	40 stars large 5 brt stars cluster
	1342	88-8	Per	7.1	OC	15	W	0331.6+3720	25 stars sl elongated poor
	1444	80-8	Per	6.4	OC	4	W	0349.4+5240	15 stars string of brt stars
*	1513	60-7	Per	8.8	OC	12	W	0410.0+4931	40 stars loose compact nucleus
*	1528	61-7	Per	6.2	OC	25	W	0415.4+5114	80 stars loosely packed rich
*	1545	85-8	Per	8	OC	18	W	0420.9+5015	25 stars loose 2 brt starswh org
	2324	38-7	Pup	8.8	OC	9	W	0704.2+0103	30 stars 5 stars = a Y loose
*	2421	67-7	Pup	9.4	OC	8	W	0736.3−2037	50 stars no shape lg fairly rich
*	2422	38-8	Pup	4.5	OC	25	W	0736.6−1430	M47
*	2423	28-7	Pup	6.9	OC	20	W	0737.1−1352	60 stars lg brt stars easy
	2438	39-4	Pup	11.3	PN	68	W	0741.8−1444	brt obvious circular within M46
	2440	64-4	Pup	11.5	PN	54 × 20	W	0741.9−1813	Greenish round easy to find
	2479	58-7	Pup	9.5	OC	8	W	0755.1−1743	40 stars faint round small easy
*	2482	10-7	Pup	8.7	OC	18	W	0754.9−2418	50 stars lg rich no shape
	2489	23-7	Pup	9.4	OC	7	W	0756.2−3004	30 stars circular 2 brt st easy
	2509	1-8	Pup	9.3	OC	4	W	0800.2−1904	40 stars some brt most faint
*	1647	8-8	Tau	6	OC		W	0446.0+1904	30 stars round double stars
	1750	43-8	Tau		OC	45	W	0503.9+2339	Part of NGC 1746 3 clusters
	1817	4-7	Tau	7.9	OC	15	W	0512.2+1642	16 stars loose scattered poor

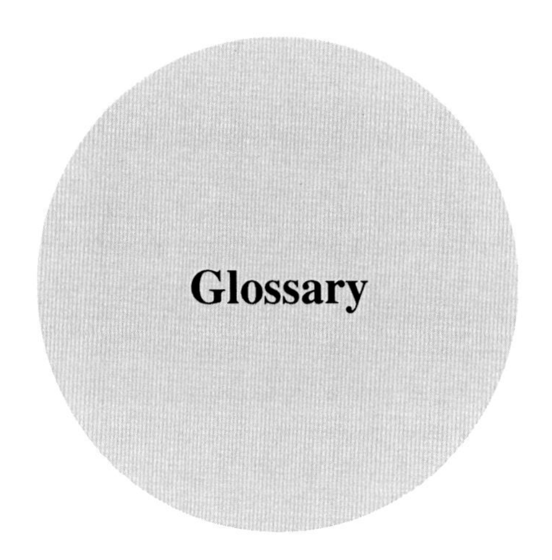

Glossary

AFOV Apparent Field of View. Usually applied to telescope eyepieces.

AltAz Altitude and Azimuth mount. Features the intuitive left–right and up–down movements of the telescope.

Antique Auction An auction offering antiques for sale from one or more individuals.

Autoguider This function processes the signal from a CCD camera installed on a guide scope, and it automatically guides the telescope and mount with high precision over an extended period. This enables long exposure photography and imaging of astronomical objects.

Backlash Compensation Provides a reduced time lag at the point of revised motion where the mount drive gears briefly lose contact.

Calcium-K Calcium K (Ca-K) telescopes and filters are used to study the wavelength of 393.4 nm. This emission line is one of two that are produced by calcium just at the edge of the visible spectrum, in a layer that is slightly lower and cooler than the layer viewed in hydrogen-alpha.

Caldwell Catalog A collection of 109 of the most impressive celestial objects culled from the NGC and IC catalogs that were not included in Messier's list. The list was created by Sir Patrick Caldwell-Moore.

Cataracts A clouding of the eye's normally clear lens.

Consignment Auction An auction offering items from multiple owners, of a mix of items of undetermined age.

Dark sky A term used to describe an area that is not affected by light pollution, and where the Milky Way is easily seen.

Diabetic retinopathy As a result of uncontrolled or poorly controlled diabetes, blood vessels in the retina are damaged, leaking fluid or blood. Fragile, brush-like

J.L. Chen, *Astronomy for Older Eyes*, The Patrick Moore Practical Astronomy Series, DOI 10.1007/978-3-319-52413-9

branches and scar tissue develop causing blurring or distortion of the images sent from the retina to the brain.

eBay An online website for buying and selling goods through a But-It-Now or online auction process.

Equatorial mount Features the ability to track an astronomical object by countering the rotation of the Earth. The RA, or right ascension, axis is set parallel to the Earth's axis. The declination axis is the axis of rotation that is at right angles to the polar axis of an equatorial mounting and that permits pointing the telescope to celestial objects of different declinations. Declination is the measurement of an objects angular distance from the celestial equator.

Estate Auction An auction offering the contents of a single individual's estate.

Estate Tag Sales These are a subset of estate sales where every item is tagged with a price, and as the tag sale continues over many days, each day brings a lower tag price for each item.

Floaters Tiny clumps of gel or cells inside the clear, gel-like fluid in the eye called the vitreous.

FOV Field-of-view. The true FOV is found by dividing the AFOV of an eyepiece by the magnification that results from using the eyepiece.

Glaucoma A disease of the optic nerve, which carries visual images from the eye to the brain.

GoTo Mount Computerized telescope mounts that automatically point the telescope towards the requested object.

Heat Index Temperature equivalency taking into account the actual temperature and the humidity.

HVAC Heating, Ventilation, and Air Conditioning

Hydrogen-Alpha H-alpha (Hα) is a specific deep-red visible spectral line created by hydrogen with a wavelength of 656.28 nm, which occurs when a hydrogen electron falls from its third to second lowest energy level. This is very useful for observing solar prominences.

IC Catalog 5386 deep sky objects cataloged by J.L.E. Dreyer as a supplement to the NGC catalog.

IDA International Dark-Sky Association, an organization seeking to reduce or eliminate light pollution. The IDA is a United States-based nonprofit organization with the mission to preserve and protect the nighttime environment and the heritage of dark skies through quality outdoor lighting.

Interocular lens implant (IOL) In cataract surgery, after the patient's original cataract-clouded lens is surgically removed, a new artificial lens IOL is implanted to take the place of the patient's original lens.

LASIK surgery Laser surgery on the eye to reshape the cornea of the eye to correct nearsightedness or farsightedness.

Light pollution The skyglow around cities and suburbs caused by unrestricted and uncontrolled use of lighting. Light pollution prevents or retricts the viewing of the night sky.

Macular degeneration The condition of the eye where damage or breakdown of the macula of the eye resulting in the gradual loss of vision.

Meridian An imaginary line drawn from due South directly overhead to due North.

Messier catalog A list of 110 (actually 109) deep sky objects created by Charles Messier in the late 1700s. It consists of 39 galaxies, 7 nebulae, 5 planetary nebulae, and 55 star clusters.

Milky Way Earth's home galaxy. Seen from Earth as an elongated star cloud across the sky.

NGC or New General Catalogue A catalog of deep sky objects based on William Herschel's Catalog of Nebulae. The NGC catalog contains 7840 objects, and was created by J.L.E. Dreyer.

Ophthalmology The branch of medicine concerned with the study and treatment of disorders and diseases of the eye.

Optometry Optometrists are trained to prescribe and fit lenses to improve vision.

Photochromic lenses Eyeglass lenses that are clear indoors or at night, and darken automatically when exposed to sunlight.

Posterior capsular opacification (PCO) When the lens capsule, the membrane that wasn't removed during surgery and supports the lens implant, becomes cloudy and impairs vision.

Presbyopia The condition that occurs after the age of 40, where the eye lens loses flexibility and focusing the eye on close-in requires reading glasses.

Public outreach Programs conducted by educational organizations and astronomy clubs to educate the general populace on the virtues of astronomy, the evils of light pollution, etc.

Scientific Instruments Auction There are occasional auction limited to just scientific instruments.

Seasonal affective disorder (SAD) A type of depression that's related to changes in seasons, with symptoms usually beginning in the fall and continuing through winter.

Sidereal rate The standard tracking rate for compensating for the Earth's rotation. This is the rate the stars move across the sky.

Skyglow A condition caused by light pollution, resulting in the obscuration of faint to bright stars in the night sky.

Star party A gathering of amateur astronomers for the purpose of observing the sky. Local star parties may be one night affairs, but larger events can last up to a week or longer and attract hundreds or even thousands of participants.

Ultraviolet (UV) Light are on the high-energy end of the visible spectrum. UV rays have higher energy than visible light rays, which makes them capable of producing changes in the skin that create a suntan. UV radiation, in moderation, has the beneficial effect of helping the body manufacture adequate amounts of vitamin D.

Unreserved Auction An auction offering goods to be sold to the highest bidder, regardless of price.

Wall wart An AC-to-DC power supply that plugs into a wall electrical socket.

YAG laser capsulotomy A laser beam makes a small opening in the clouded capsule to allow light through.

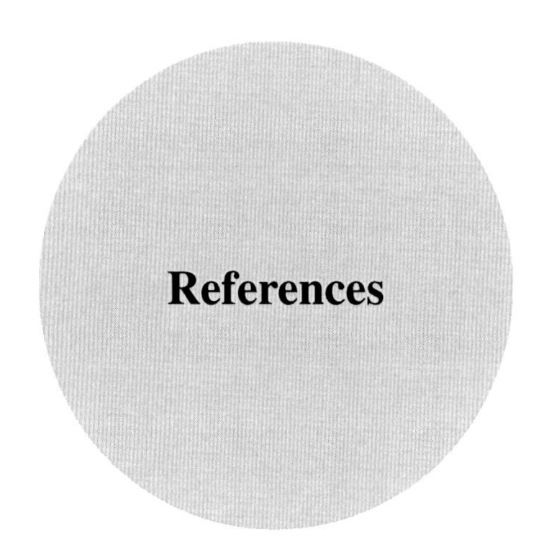

References

Books

Balch, Phyllis A., CNC, *Prescription for Herbal Healing*, Avery, Penguin Putnam Inc., 2002.

Dick, Steven J., Lupisella, Mark L. editors, *Cosmos & Culture,* National Aeronautics and Space Administration Publication, 2009.

Dickinson, Terrence, Dyer, Alan, *The Backyard Astronomer's Guide,* Third edition, Firefly Books Ltd., 2008.

English, Neil, *Classic Telescopes,* Springer, 2013.

Fulton, Ken, *The Light-Hearted Astronomer*, AstroMedia, 1984.

Hagen, Phillip, MD editor, Mayo Clinic Guide to Self-Care, Mayo Clinic Health Information, 2003.

Harrington, Phillip S., Star Ware, 4th edition, John Wiley & Sons Inc., 2007.

Kier, Ruben, *The 100 Best Astrophotography Targets,* Springer, 2009.

Sidgwick, J.B., Amateur Astronomer's Handbook, 3rd edition, Dover Publications, 1971.

Internet Source

"45 Manly Hobbies", The Art of Manliness, Retrieved May 2016, http://www.artofmanliness.com/2010/01/06/45-manly-hobbies/

"7 Best Foods for Your Eyes", Cooking Light, Retrieved May and June 2016, http://www.cookinglight.com/eating-smart/nutrition-101/foods-for-eyes/tips-to-keep-eyes-healthy

"10 Reasons You Need to Teach Your Homeschooler Astronomy", Intoxicated on Life, Retrieved May and June 2016, http://www.intoxicatedonlife.com/2014/07/31/reasons-teach-astronomy/

"Activity Theory." Boundless Sociology. Boundless, 26 May, 2016. Retrieved 31 May, 2016 from https://www.boundless.com/sociology/textbooks/boundless-sociology-textbook/aging-18/the-functionalist-perspective-on-aging-128/activity-theory-722-9146/

J.L. Chen, *Astronomy for Older Eyes*, The Patrick Moore Practical Astronomy Series, DOI 10.1007/978-3-319-52413-9

"The AAVSO 2011 Demographic and Background Survey", Retrieved May 2016, http://arxiv.org/pdf/1204.3582.pdf
"Aging and Your Eyes", National Institute for Aging, Retrieved June 2016, https://www.nia.nih.gov/health/publication/aging-and-your-eyes
"The Aging Eye: See into Your Future", Share Care, Retrieved May 2016, https://www.sharecare.com/health/eye-vision-health/article/aging-eye
"Adult Vision: Over 60 Years of Age", American Optometric Association, Retrieved May and June 2016, http://www.aoa.org/patients-and-public/good-vision-throughout-life/adult-vision-19-to-40-years-of-age/adult-vision-over-60-years-of-age?sso=y
"Astronomical Observing with Binoculars", Stargazing.net, Retrieved June 2016, http://www.stargazing.net/David/binoculars/index.html
"Astronomical Optics", Retrieved June 2016, http://www.handprint.com/ASTRO/ae3.html
"Astronomy as a Hobby for Seniors & Disabled", Disabled World, Retrieved June 2016, http://www.disabled-world.com/entertainment/hobby/astronomy/
"Astronomy in the starry desert", Tierra Hotels, Retrieved June 2016, http://www.tierrahotels.com/tierra-atacama-hotel-boutique-amp-spa/excursions/stargazing.htm
"The Best Foods for Healthy Eyes", Mercola, Retrieved June 2016, http://articles.mercola.com/sites/articles/archive/2015/08/03/best-foods-for-eye-health.aspx
"Big Island Stargazing", Star Gaze Hawaii, Retrieved June 2016, https://www.stargazehawaii.com/stargazing/
"Bino-viewers", One Minute Astronomers, Retrieved June 2016, http://oneminuteastronomer.com/4940/binoviewers/
"Cataract Surgery", Mayo Clinic, Retrieved May and June 2016, http://www.mayoclinic.org/tests-procedures/cataract-surgery/multimedia/cataract-surgery/img-20005853
"Choices for the Amateur", Retrieved May 2016, http://www.quadibloc.com/science/opt0201.htm
"Choosing and Using Binoculars", Amateur Astronomers of New York, Retrieved June 2016, http://www.aaa.org/articles/choosing-and-using-binoculars/
"Considerations for Older Travelers", U.S. State Department, Retrieved June 2016, https://travel.state.gov/content/passports/en/go/older-traveler.html
"The Effects of Aperture Obstruction", Telescope Optics.net, Retrieved June 2016, http://www.telescope-optics.net/obstruction.htm
"Exercise: Benefits of Exercise", NIH Senior Health, Retrieved June 2016, https://nihseniorhealth.gov/exerciseforolderadults/healthbenefits/01.html
"Exercise Recommendations for Older Adults" by LaVona S. Traywick, PhD, Today's Geriatric Medicine, Retrieved June 2016, http://www.todaysgeriatricmedicine.com/news/ex_092210_03.shtml
"Eye Strain in the Elderly", Senior Health 365, Retrieved June 2016, http://www.seniorhealth365.com/2012/02/08/eye-strain-in-the-elderly/?fdx_switcher=true
"Four Fantastic Foods to Keep Your Eyes Healthy", American Academy of Ophthalmology, Retrieved May and June 2016, http://www.aao.org/eye-health/news/four-fantastic-foods
"Good Foods for Eye Health", WebMD, Retrieved May and June 2016, http://www.webmd.com/healthy-aging/nutrition-world-3/foods-eye-health?page=2
"How much physical activity do older adults need?", Centers for Disease Control and Prevention, Retrieved June 2016, http://www.cdc.gov/physicalactivity/basics/older_adults/index.htm
"How Stress Affects Your Health", American Psychological Association, Retrieved June 2016, http://www.apa.org/helpcenter/stress.aspx
"How will my prior LASIK procedure affect cataract surgery?", American Academy of Ophthalmology, Retrieved May and June 2016, http://www.aao.org/eye-health/ask-ophthalmologist-q/how-does-previous-lasik-affect-cataract-surgery
"How Your Vision Changes As You Age" by Gary Heiting OD, All About Vision.com, retrieved May and June 2016, http://www.allaboutvision.com/over60/vision-changes.htm
"LASIK with Cataracts", American Academy of Ophthalmology, Retrieved June 2016, http://www.aao.org/eye-health/ask-ophthalmologist-q/lasik-with-cataracts

"The Life Expectancy of Astronomers", http://people.physics.tamu.edu/krisciunas/astrs.html, Retrieved May, 2016, http://www.ncbi.nlm.nih.gov/pmc/articles/PMC3940510/
"Light Grasp of a Telescope", Starizona, Retrieved June 2016, https://starizona.com/acb/basics/observing_theory.aspx
"Observing Field Etiquette", Western Kentucky Amateur Astronomers, Retrieved June 2016, http://wkaa.net/etiquette.php#.V36tycefEUR
"Observing Programs Arranged by Experience Level of Observer", The Astronomical League, Retrieved June 2016, https://www.astroleague.org/al/obsclubs/LevelObservingClubs.html
"Observing the sun in Ca II K", Stephan Ramsden Solar Scope Reviews, Retrieved June 2016, http://www.stephenramsden.com/solarastrophotography/Observing%20in%20Ca%20II%20K.pdf
"Photochromic Lenses: Transitions Lenses and Other Light-Adaptive Lenses", All About Vision, Retrieved June 2016, http://www.allaboutvision.com/lenses/photochromic.htm
"Physical Activity and Older Adults", World Health Organization, Retrieved June 2016, http://www.who.int/dietphysicalactivity/factsheet_olderadults/en/
"Retirement Hobbies", Power to Change, Retrieved May 2016, http://powertochange.com/culture/hobbyideas/
"Seven Safety Tips for Senior Travelers", Independent Traveler.com, Retrieved June 2016, http://www.independenttraveler.com/travel-tips/senior-travel/seven-safety-tips-for-senior-travelers
"Solar Scope Reviews", Stephan Ramsden Solar Scope Reviews, Retrieved June 2016, http://www.solarscopereviews.com
"Stargazers keep on looking up", The Baltimore Sun, Retrieved May, 2016, http://articles.baltimoresun.com/2007-10-10/news/0710100111_1_astronomy-clubs-telescopes-amateur-astronomers
"Strength and muscle mass loss with aging process. Age and strength loss", Muscle, Ligiments, and Tendons Journal, Retrieved May and June 2016, http://www.dummies.com/how-to/content/what-happens-to-aging-muscles.html
"Telescope Performance Factors Central Obstruction and Wave Error" by Paul Laughton, The Dr. Richard Feynman Observatory, Retrieved May and June 2016, http://www.laughton.com/paul/rfo/obs/obs.html
"Top 6 tips for using ordinary binoculars for stargazing", Earth Sky News, Retrieved June 2016, http://earthsky.org/astronomy-essentials/top-tips-for-using-ordinary-binoculars-for-stargazing
"Travel tips for seniors", Better Health Channel, Retrieved June 2016, https://www.betterhealth.vic.gov.au/health/healthyliving/travel-tips-for-seniors
"Ultraviolet (UV) Radiation And Your Eyes", All About Vision, Retrieved June 2016, http://www.allaboutvision.com/sunglasses/spf.htm
"Vision and Eye Problems in Aging Adults", Web MD, Retrieved June 2016, http://www.webmd.com/eye-health/vision-problems-aging-adults
"What Happens to Aging Muscles", Agin, Brent and Perkins, Sharon, *Healthy Aging for Dummies,* Retrieved May and June 2016
"What Are The Effects Of The Central Obstruction?", Retrieved June 2016, http://www.hoflink.com/~mkozma/obstruction.html
"What is Estate Planning?", Estate Planning.com, Retrieved June 2016, https://www.estateplanning.com/What-is-Estate-Planning/
"Why does eyesight deteriorate with age?" bu David Zacks, Scientific American, Retrieved June 2016, http://www.scientificamerican.com/article/why-does-eyesight-deterio/
"Why Observe Occultations?", International Occultation Timing Association, Retrieved May 2016, http://occultations.org/occultations/why-observe-occultations

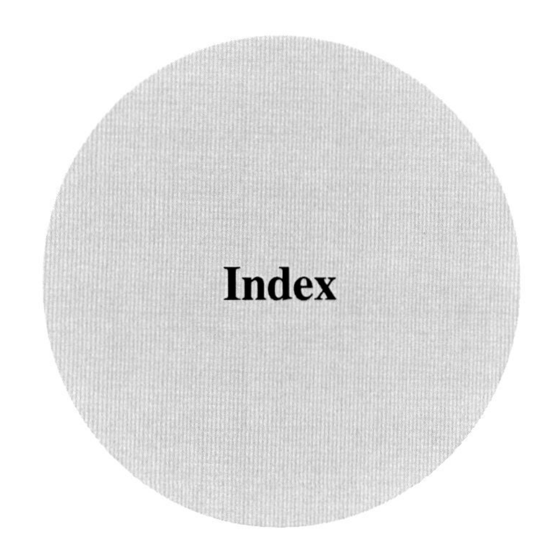

Index

J.L. Chen, *Astronomy for Older Eyes*, The Patrick Moore Practical Astronomy Series, DOI 10.1007/978-3-319-52413-9